Netscape Communicator

FOR BUSY PEOPLE

Blueprints for Netscape Communicator

On the following pages, we provide blueprints for some of the best ways to use Netscape Communicator:

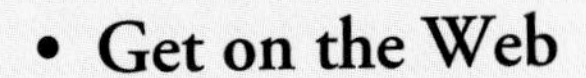

- Get on the Web
- Search the Web
- Get Instant Web Delivery
- Use Powerful Internet E-mail
- Discuss and Communicate with Internet Newsgroups
- Collaborate with Internet Conferencing
- Create Simple Web Pages
- Use Netscape's Wizards for Instant Web Pages
- Develop Complex Web Pages and Forms

BLUEPRINT FOR BUSY PEOPLE™

Get on the Web

Use Netscape Navigator's toolbar to move among Web sites (pages 57-61).

Search the Web for information, people, and more (pages 72-78).

Navigate the alphabet soup of URLs, Web addresses, and Web resources (pages 53-57).

Jump right into the Web—without any hassle (pages 19-26).

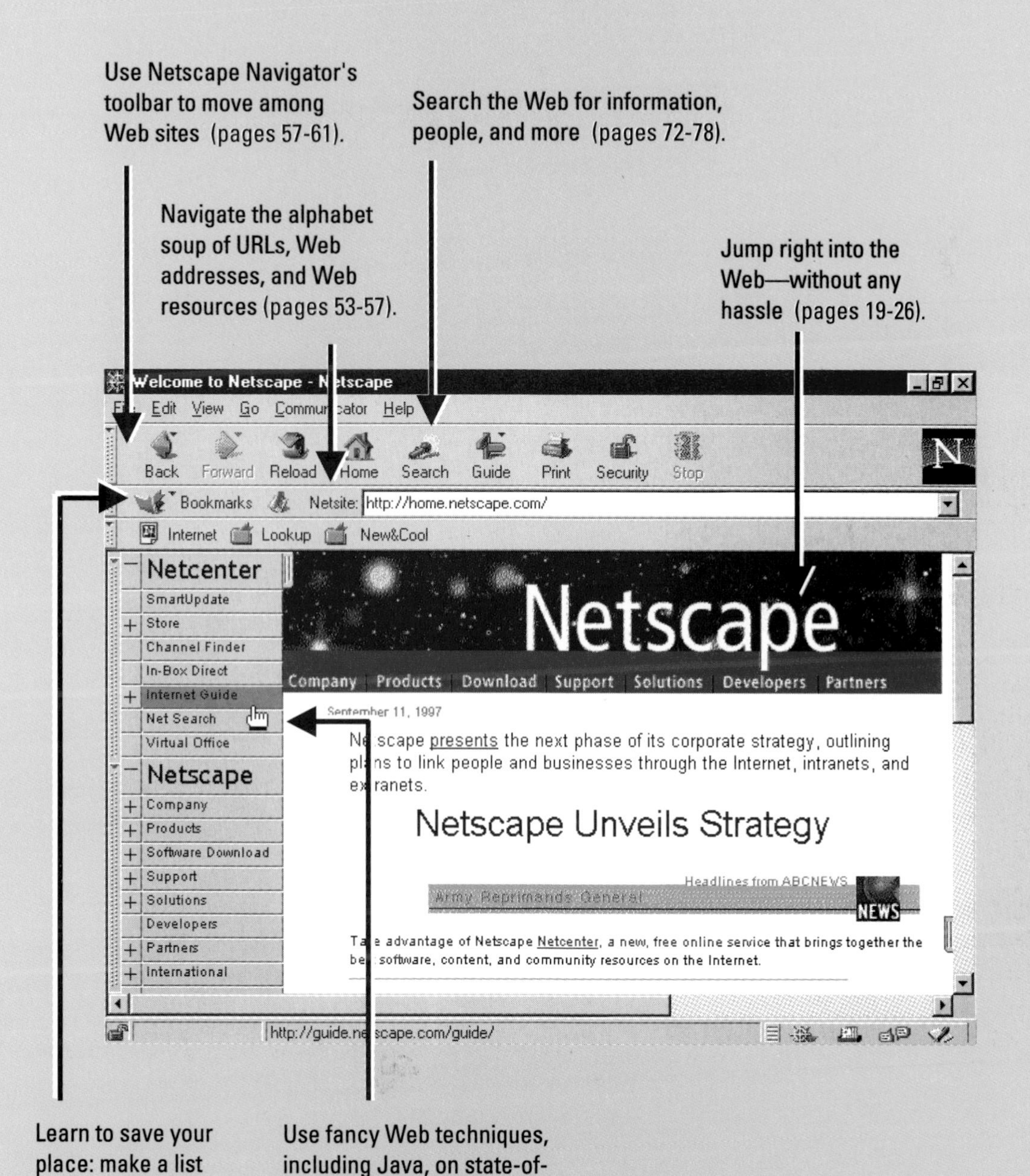

Learn to save your place: make a list of your favorites (pages 63-66).

Use fancy Web techniques, including Java, on state-of-the-art sites (page 26).

Search the Web

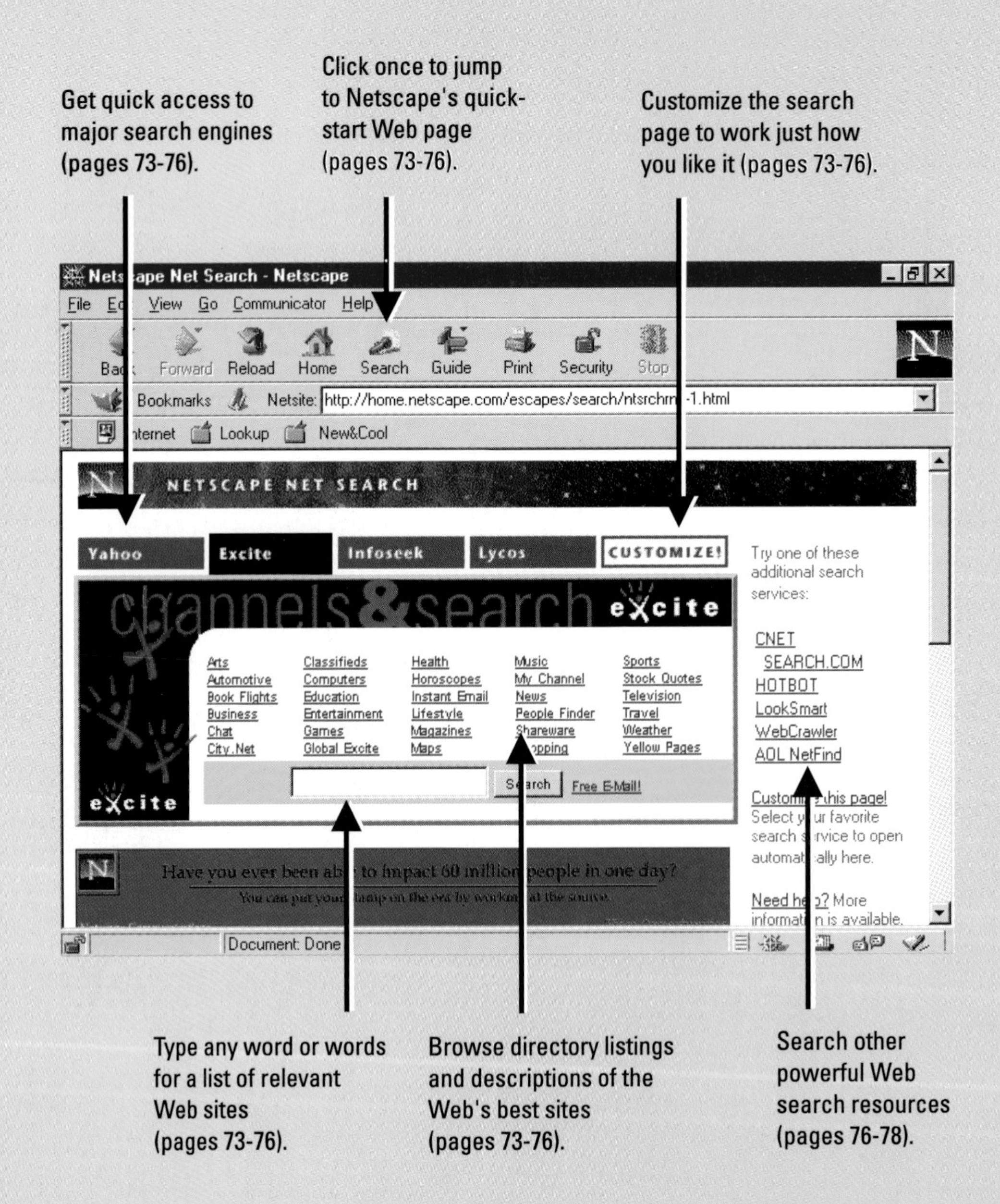

BLUEPRINT FOR BUSY PEOPLE™

Get Instant Web Delivery

Get Web sites delivered to your desktop automatically (pages 93-95).

Add the hottest Web "channels" in an instant (pages 99-102).

Hide the full control panel to view the whole Netcaster screen (pages 102-104).

Choose from scores more Netcaster channels (pages 98-99).

Control your Netcaster Webtop—or switch to a regular Netscape Navigator window (pages 102-104).

BLUEPRINT FOR BUSY PEOPLE™

Use Powerful Internet E-mail

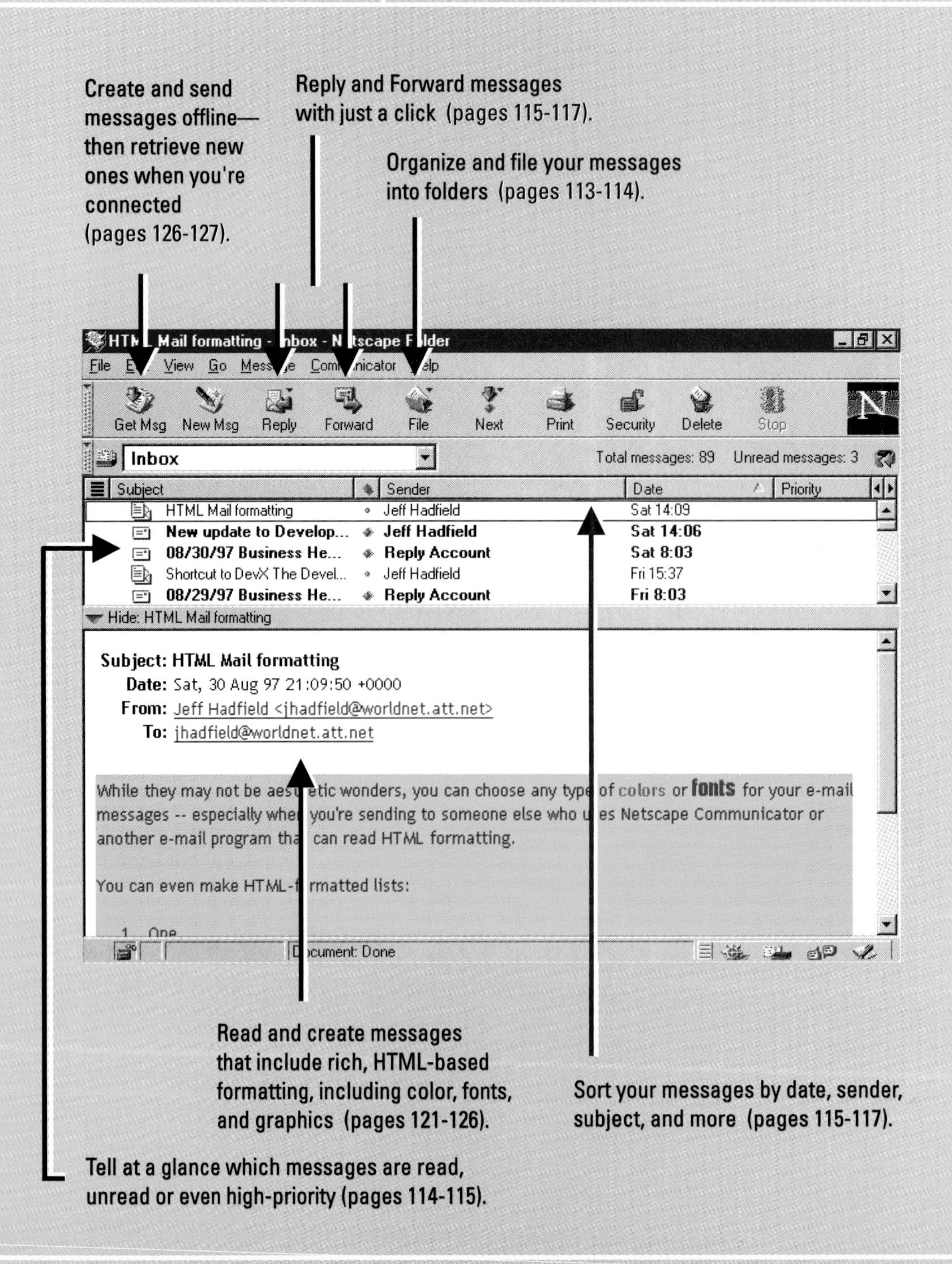

BLUEPRINT FOR BUSY PEOPLE™

Discuss and Communicate with Internet Newsgroups

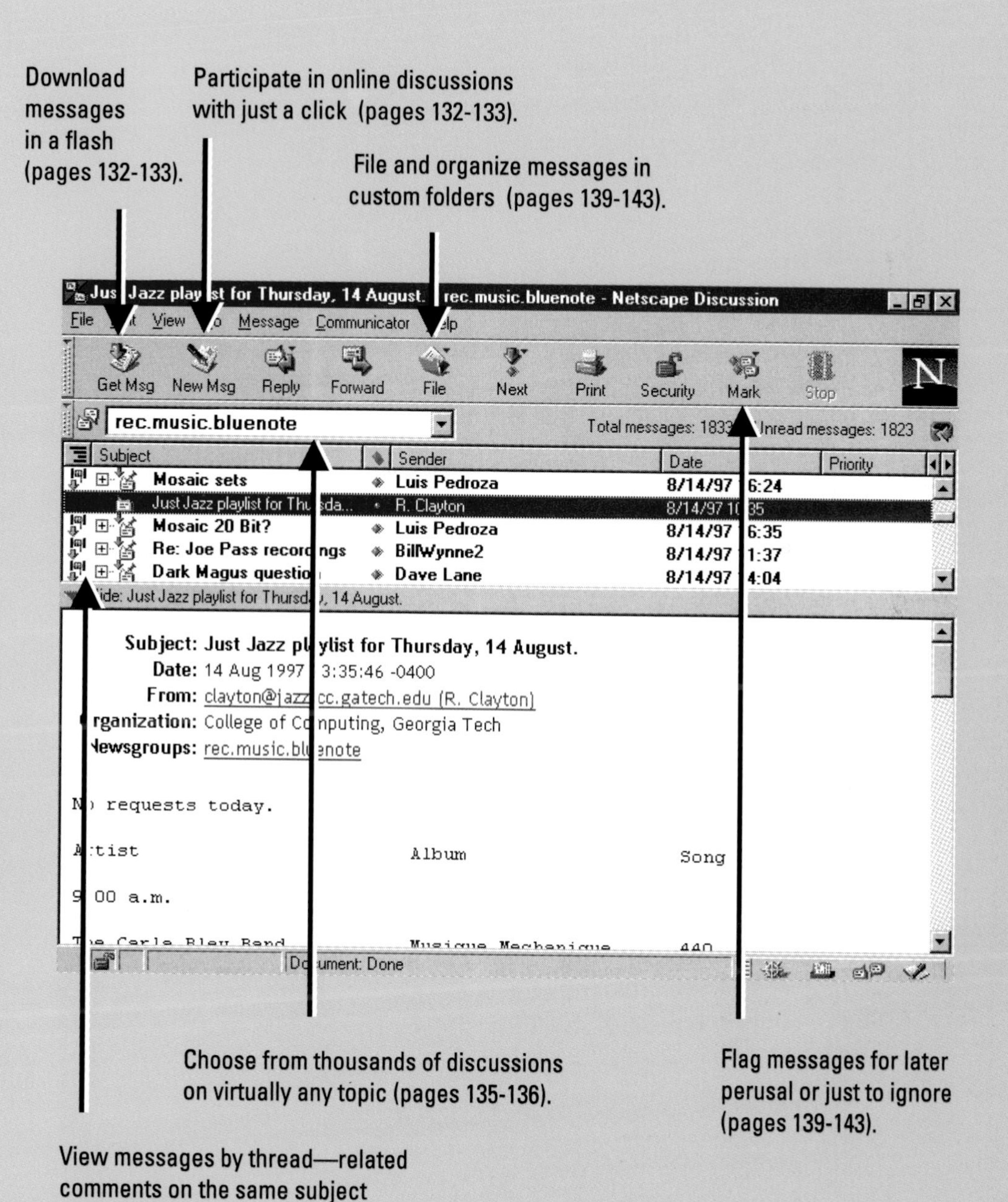

BLUEPRINT FOR BUSY PEOPLE™

Collaborate with Internet Conferencing

Collaborate using a shared whiteboard (pages 156-158).

Browse the Web together—while you comment on the sites you're seeing (page 158).

Send and receive files with your collaborator (pages 158-159).

Chat "live" using a sophisticated, text-based utility (pages 159-160).

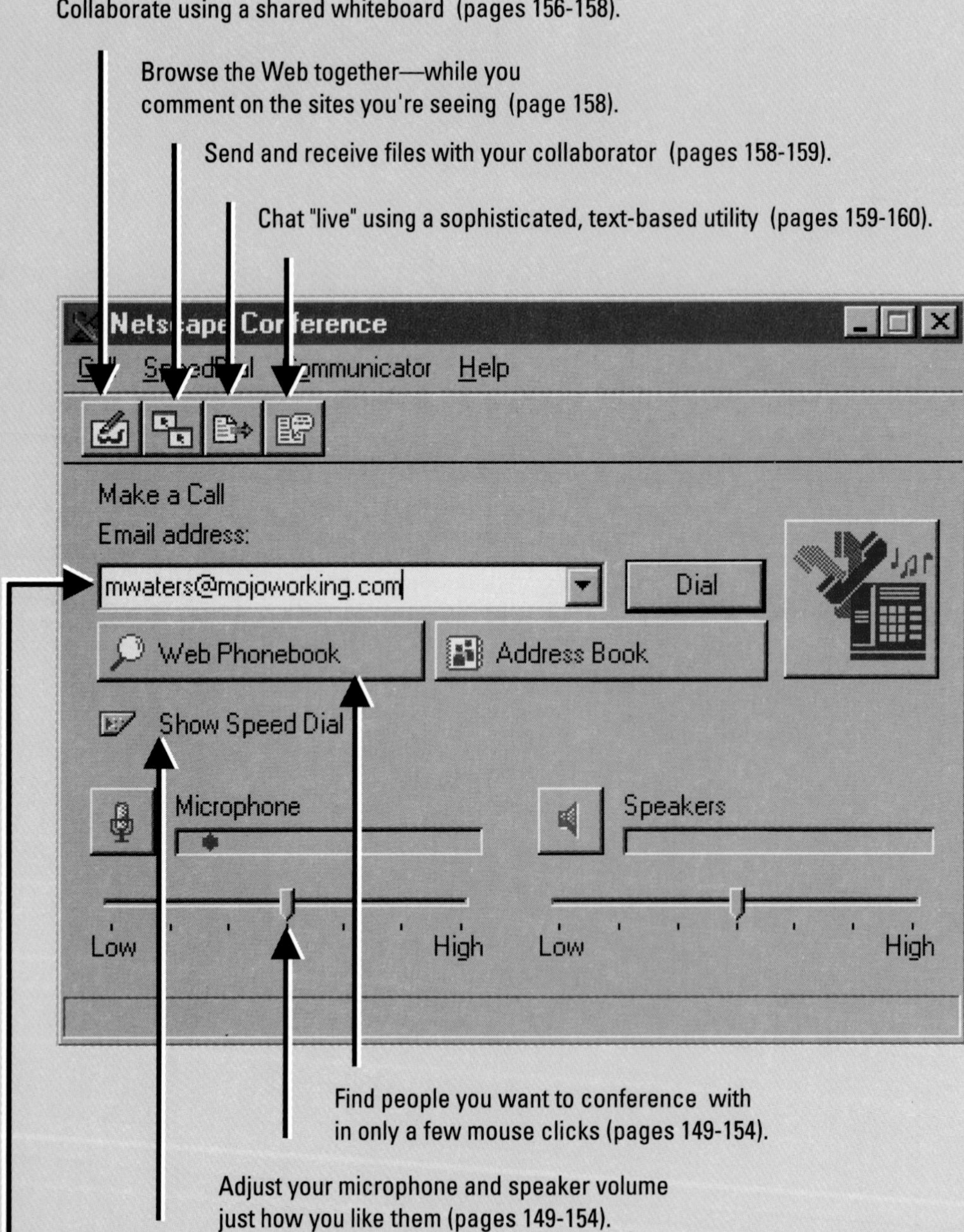

Find people you want to conference with in only a few mouse clicks (pages 149-154).

Adjust your microphone and speaker volume just how you like them (pages 149-154).

Place frequently called people into your speed dial list (pages 154-156).

Place a conference call to anyone with Netscape Communicator 4.0—using just an e-mail address (pages 154-156).

BLUEPRINT FOR BUSY PEOPLE™

Create Simple Web Pages

Give your Web page a title—that appears in the Web browser just like this (pages 197-198).

Use HTML styles—like headlines—to add rich formatting to your pages (pages 200-203).

Creating Web pages with Netscape Composer is only a click away (page 176).

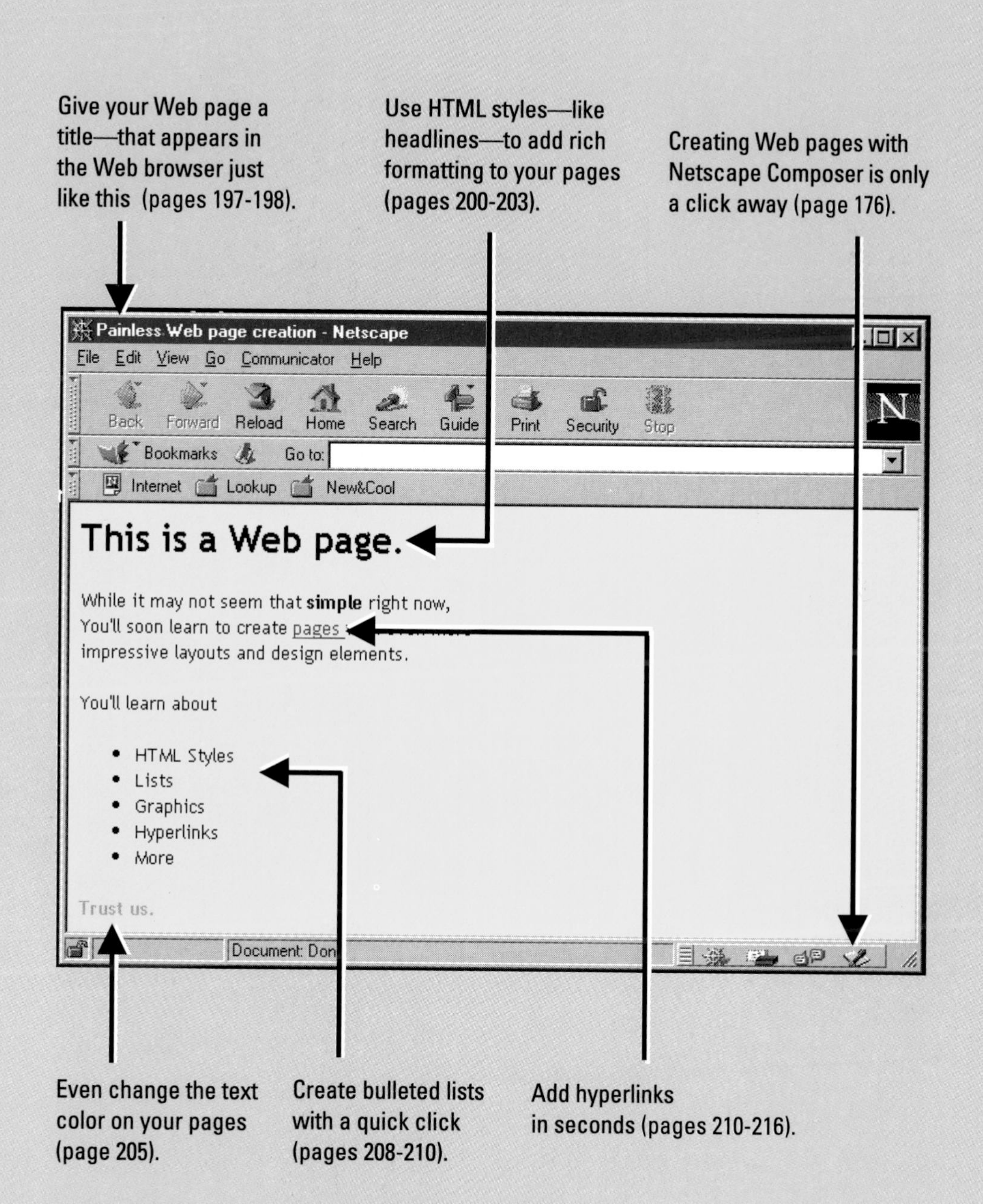

Even change the text color on your pages (page 205).

Create bulleted lists with a quick click (pages 208-210).

Add hyperlinks in seconds (pages 210-216).

BLUEPRINT FOR BUSY PEOPLE™

Use Netscape's Wizards for Instant Web Pages

Netscape Composer makes Netscape's Wizards a quick menu choice away (pages 178-184).

See an instant preview of your finished Web page here (pages 178-184).

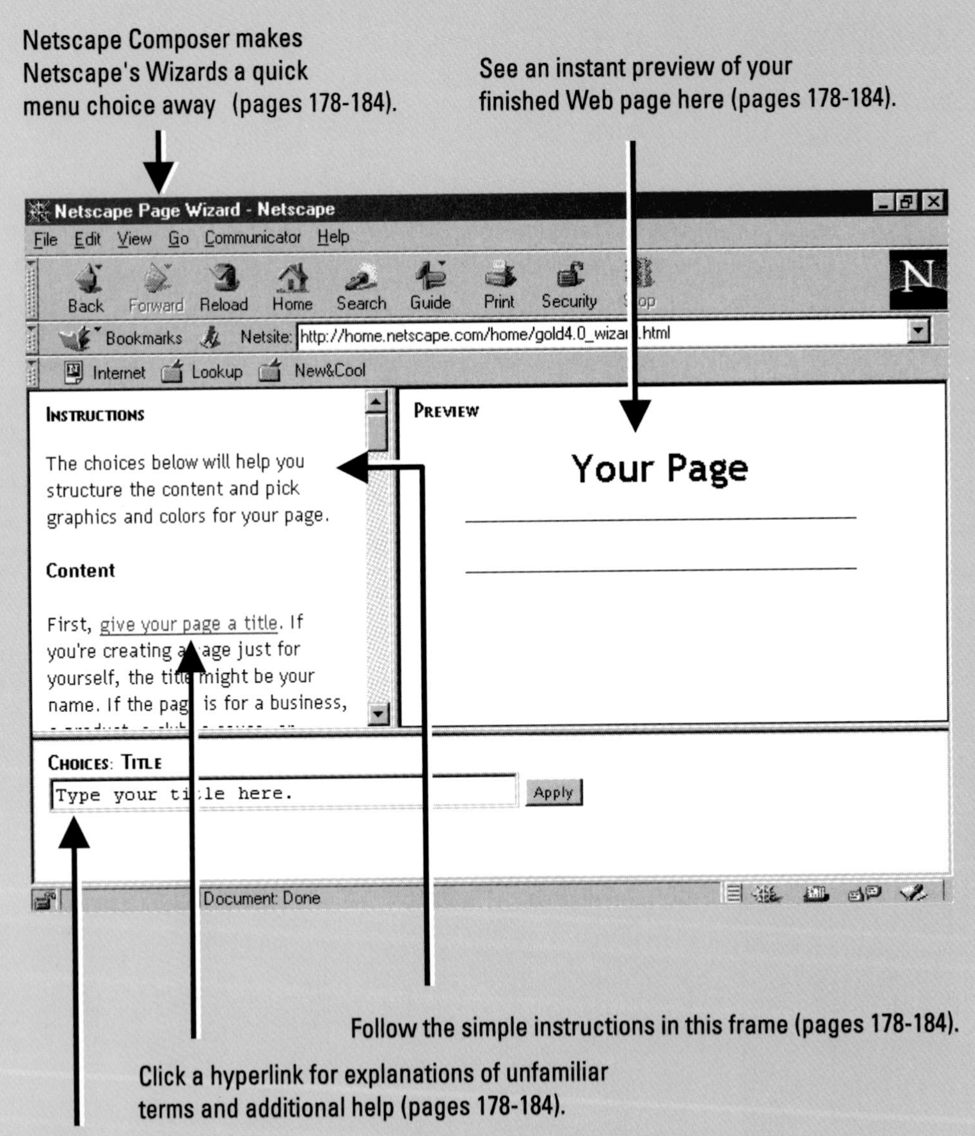

Follow the simple instructions in this frame (pages 178-184).

Click a hyperlink for explanations of unfamiliar terms and additional help (pages 178-184).

Type your own text and material in this frame (pages 178-184).

BLUEPRINT FOR BUSY PEOPLE™

Develop Complex Web Pages and Forms

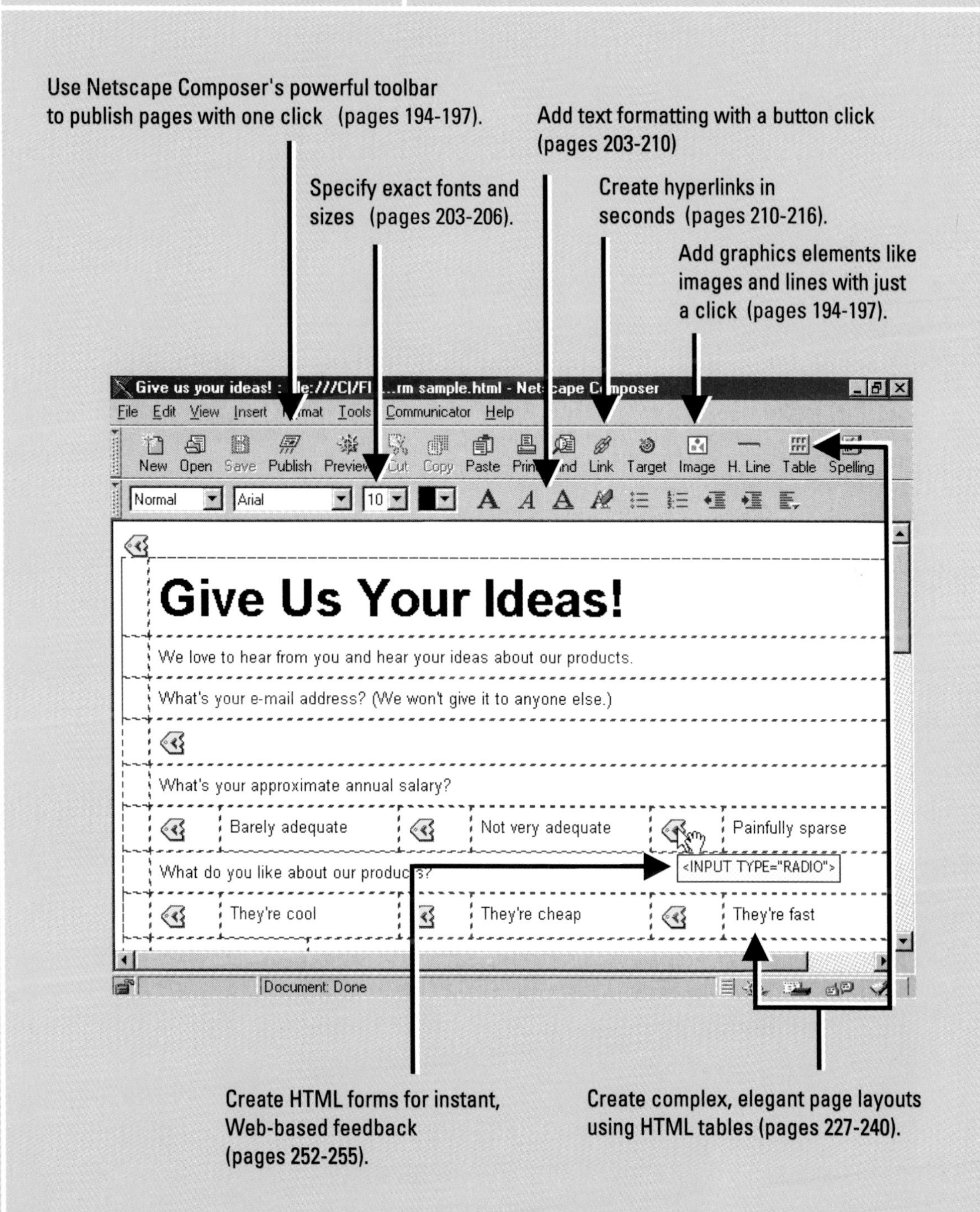

Netscape Communicator

FOR BUSY PEOPLE

The Book to Use When There's No Time to Lose!

Christian Crumlish
Jeff Hadfield

OSBORNE

Osborne/**McGraw-Hill**

Berkeley / New York / St. Louis / San Francisco / Auckland / Bogotá
Hamburg / London / Madrid / Mexico City / Milan / Montreal / New Delhi
Panama City / Paris / São Paulo / Singapore / Sydney / Tokyo / Toronto

A Division of The **McGraw-Hill** *Companies*

Osborne/**McGraw-Hill**
2600 Tenth Street
Berkeley, California 94710
U.S.A.

For information on translations or book distributors outside the U.S.A., or to arrange bulk purchase discounts for sales promotions, premiums, or fund-raisers, please contact Osborne/**McGraw-Hill** at the above address.

Netscape Communicator for Busy People

1234567890 DOC DOC 901987654321098

ISBN 0-07-882441-9

Publisher: Brandon Nordin
Acquisitions Editor: Joanne Cuthbertson
Editor-in-Chief: Scott Rogers
Acquisitions Editor: Joanne Cuthbertson
Project Editor: Claire Splan
Editorial Assistant: Gordon Hurd
Technical Editor: Allen Biehl
Copy Editor: Sally Engelfried
Proofreader: Joe Sadusky
Indexer: David Heiret
Computer Designers: Roberta Steele, Mickey Galicia, Peter F. Hancik
Series and Cover Designer: Ted Mader Associates
Series Illustrator: Daniel Barbeau

For Jennifer, Nick and Hayley

&

SBN

About the Authors

Christian Crumlish, publisher of the multimedia magazine *Enterzone*, is a writer of computer books, stories, and hyperfiction. He is the author of *The Internet for Busy People* and *Word for Windows 95 for Busy People.*

Jeff Hadfield has spent years making technical subjects understandable. He's the former editorial director of the WordPerfect magazines, and author of *Web Publishing with Corel WordPerfect Suite 8* (Osborne/McGraw-Hill). He's produced electronic publications and Web sites, as well as written and presented on the creation of effective Web sites to the magazine industry. He's currently editor in chief of *Visual Basic Programmer's Journal*, a Fawcette Technical Publication. He has worked with personal computers for almost 20 years.

Contents at a glance

Contents

PRIORITY!
MEMO:
FIRST CLASS
OVERNIGHT
RUSH!

Acknowledgments

We were luckier than most. Many times, computer books must be written before the software's even released—but we were able to work on this project after Netscape Communicator 4.0 was closer to final. For you, that means that you're reading a book that more accurately reflects the software you'll actually be using. For us, it meant that the software wasn't changing as quickly as we wrote about it. Of course, the Internet moves pretty fast—so we tried to cover the fundamentals of what you need to know, regardless of what tomorrow brings.

We were also lucky to work with an excellent team. Thanks to Joanne Cuthbertson for holding our feet to the fire and our noses to the grindstone. Ted Mader's witty, sophisticated design, built around Dan Barbeau's frazzled, contemporary people-oids, set the tone for the project and kept our minds firmly focused on you, the kind of person who needs a book like this.

Our technical editor, the invaluable Allen Biehl, made this book much more than it could have been without his contributions. His dead-on, in-the-trenches commentary and corrections brought you the best Netscape Communicator book possible. And he did it all without complaining, at least, not to us.

Thanks also to Malcolm Humes, the co-author of the original edition of this book, *Web Publishing with Netscape Navigator [Gold] for Busy People*. His original efforts live on in this edition.

Claire Splan, our Project Editor, kept us in line and made our how-to information even more valuable for you. Our copyeditor, Sally Engelfried, made us sound sensible and reasonable—and helped us blend our styles.

The art and production team, headed by Marcela Hancik and including Roberta Steele, Lance Ravella, Jani Beckwith, Mickey Galicia, Sylvia Brown, and Peter F. Hancik, produced beautiful galleys that required a minimum of correction and polishing before they turned into final pages.

We're also grateful to those who helped us compile the unique list of Web sites found in Appendix A: E. Stephen Mack, Jennifer Hadfield, and the Antiweb gang, including Garrett Keogh, Phil Franks, and others.

Jeff thanks his patient wife, Jennifer, and untiring distractions Nick and Hayley, for their love and support. Thanks also to his parents—both Hadfields and Stringfellows—for their hospitality and even babysitting during the move back to California and this project.

Christian, if it's permissible, thanks *Jeff* foremost, for taking the lead on this project, and Osborne's elite editorial and production crews for their impeccable work.

Introduction

With the Internet and the rest of the computer world changing so rapidly, it's almost impossible to keep up, even if you work in this field! On top of that, with the overall "downsizing" crunch that's been rippling through the world economy, most of us are doing what used to be considered two or three different jobs, all at once. Who can blame you if you haven't mastered the latest version of Microsoft this or Netscape that? We've written this book with busy people in mind—people who know how to operate their computers but don't have the time or energy to research and compare all the various sources of information on the latest thing.

Instead, we've filtered and boiled down our expertise and experience to serve you up a concentrated and to-the-point briefing. You'll learn only what you need to know about getting the most from Netscape Communicator to use Net information and publish on the Web.

You can spend a lot of time learning technical arcana about how the Web works. Frankly, you don't need to know how it works to use it. You don't need to know how your car works to drive it—and Netscape Communicator's the same way. While you may find it useful to know a bit more when troubleshooting, for the most part, you'll never need that information. And, besides, unlike a car, software can't leave you stranded if it breaks down—you can always re-start. Likewise, there are myriad ways to create Web documents and weave together Web sites, so we've chosen an approach in the last section of the book that takes most of the arcane details out of your hands: using Netscape Communicator's Composer module. We think that a busy person with plenty else to do besides type a lot of HTML code will appreciate a handy tool that requires little more than knowing how to work a Web browser and how to use a generic word processing program.

We Know You're in a Hurry, So...

Let's agree to dispense with the traditional computer book preliminaries. You've probably used a mouse, held down two keys at once, and have heard of this vast global network called the Internet. Ideally, you've browsed around the Web a little yourself. (You wouldn't try writing a book if you'd never read one, would you?)

The first chapter will introduce you to Netscape Communicator in its various incarnations. You'll learn how to find it, buy it, download it, and install it. In Chapter 2 you'll learn how to get everything set up for the first time—stuff you'll probably never have to mess with again.

You'll probably find Chapters 3 and 4 essential: Chapter 3 tells you about Netscape Navigator: your gateway to the Web. Chapter 4 helps you find things: information, software, and people.

Much buzz has, um, buzzed around the computer industry about "broadcasting" over the Internet: Chapter 5 introduces you to Netscape's broadcast component, Netscape Netcaster—something like a TV tuner for Web sites. Chapter 6 brings you back to ground with an essential Internet function: sending and receiving e-mail with Netscape Messenger.

While e-mail helps you send messages to others, two other Netscape Communicator components help you collaborate with others. Chapter 7 introduces how to use Netscape Messenger for Internet discussions and newsgroups. Chapter 8 helps you conduct virtual meetings online with Netscape Conference.

The remainder of the book—chapters 9 through 12—gives you a quick tour through Netscape Composer. You'll learn the basics of creating Web sites and pages with this easy, pushbutton tool. (As an added bonus to the book, we've added a list of favorite and useful Web sites in Appendix A.)

Accessing the Internet With Windows 95 or Other Computers

This book uses examples and illustrations showing the Windows 95 versions of Netscape Communicator 4.0, and other programs and features, but all of the information in the book applies equally well to other types of computers and operating systems, including earlier versions of Windows, Macintosh, and UNIX systems.

Things You Might Want to Know About This Book

You can read this book more or less in any order. We suggest cruising Chapter 1 and reading Chapter 2 first, but you can start just as easily with Chapter 4 or by jumping ahead to whatever topics are new or interesting for you. Use the book as a reference. When you're stuck, not sure how to do something, know there's an answer but not what it is, pick up the book, zero in on the answer to your question, and put the book down again. Besides clear, coherent explanations of the ins and outs of Netscape Communicator 4.0, you'll also find some special elements to help you get the most out of Netscape. Here's a quick run down of the other elements in this book.

Blueprints

Want to figure out what you want quickly? Look at the front of the book for the Blueprints. They help you identify what you want to learn by pointing at it in a picture—as in "I just want to do *that.*" You don't have to know the name of what you want to do—only what it looks like.

Fast Forward

Each chapter begins with a section called Fast Forward. They should always be your first stop if you are a confident user, or impatient or habitually late. You might find everything you need to get back in stride. Written step-by-step, point-by-point, Fast Forwards are, in effect, a book within a book, a built-in quick reference guide, summarizing the key tasks explained in each chapter. This shorthand may leave you hungry, especially if you are new to the Internet. So, for more complete and leisurely explanations of techniques and shortcuts, read the rest of the chapter or follow the page references to get more complete information on a particular feature or task.

Expert Advice

Expert Advice is just that: detailed help from the experts. You'll learn time-saving tips, techniques, and worthwhile addictions. These sidebars provide perspective through asides, glimpses of the big picture, and help planning ahead.

Shortcuts

We'll always show you the easy way to get things done. But sometimes there's a faster way—maybe it's harder to remember, but it's a more direct route to get where you're going. Look for these Shortcuts sprinkled throughout the book.

Cautions

Sometimes it's too easy to plunge ahead and fall down a rabbit hole, resulting in hours of extra work just to get you back to where you were before you went astray. This icon will warn you before you commit time-consuming mistakes.

Definitions

Usually, we explain computer or networking jargon in the text, wherever the technobabble first occurs. But if you encounter words you don't recognize, look for this body builder in the margin. Definitions point out important terms you might not know the meaning of. When necessary, they're strict and a little technical, but most of the time they're informal and conversational.

Step by Steps

Blue Step by Step boxes walk you through some essential procedures, providing straightforward, 1-2-3 instructions.

Is This for Me?

We said the book's written for busy people, and we weren't kidding. Look for special notes marked "Is This for Me?" in each chapter. They'll help you decide whether you need to read the chapter or section—no point in reading about something you'll never use.

One last note: World Wide Web addresses, also called URLs, are notoriously long and strangely punctuated. Often, a Web address will not fit on a single line of text. To avoid introducing spurious characters that will make the addresses actually incorrect, Web addresses are wrapped without hyphens or any other special characters added, usually after a slash (/) or dot (.) character. So, for example, to point your Web browser at **http://ezone.org/ez/e7/articles/xian/spam.html**, just type the entire address on one line without any spaces or breaks (and don't type the comma at the end—that's just part of this sentence).

Let's Do It!

Ready? Let's dive in! Incidentally, we're happy to hear your reactions, feedback, or even corrections to this book. You can reach us through the publisher or on the Net (**hadfield@iname.com** and **xian@pobox.com**).

CHAPTER

1

Diving Headfirst into the Web

INCLUDES

- Downloading Netscape Communicator
- Choosing the best version for your needs
- Installing the software
- Running the program
- Learning the basics
- Understanding Netscape's vision

GET THE LATEST NETSCAPE SOFTWARE

Tune Up to Communicator
Get Any Netscape Software
For Subscribers Only

Download the Software ➤ pp. 4-10

1. Specify the version of Netscape Communicator 4.0 you want to download.
2. Specify your computer's operating system (Windows 95, Macintosh, etc.).
3. Specify from where you want to download the software (you can pick from several sites—choose the one geographically closest to you).
4. Specify where you want to save the installation file (save it somewhere you won't lose it—like on your Windows desktop or in your My Documents folder).

Install Netscape Communicator 4.0 ➤ pp. 11-19

Select the type of setup you prefer.

Typical — Program will be installed with the most common options. Recommended for most users.

1. Run the installation program you downloaded.
2. If necessary, close all other programs (use ALT-TAB to move between them) and click on Next to proceed.
3. Read the Software License Agreement and, assuming you agree to the terms, click Yes.
4. If you want to install Netscape Communicator 4.0's most commonly used components, choose Typical.
5. If you want to change the Destination Directory folder, click Browse and select a different folder (you can create a new one if you like). Click Next.
6. If prompted, click Next to allow Netscape to open Internet-related files by default.
7. Click Next to place the programs in a Netscape Communicator program folder.
8. Review your settings, then choose Install to install the software.

Launch Netscape Communicator 4.0 from the Desktop ➤ *pp. 19-20*

Double-click the Netscape Communicator 4.0 icon.

Launch Netscape Communicator 4.0 from the Windows 95 Start Menu ➤ *pp. 19-20*

1. Choose Start | Programs | Netscape Communicator.
2. Choose the module you want: Navigator, Composer, Collabra, or Messenger.

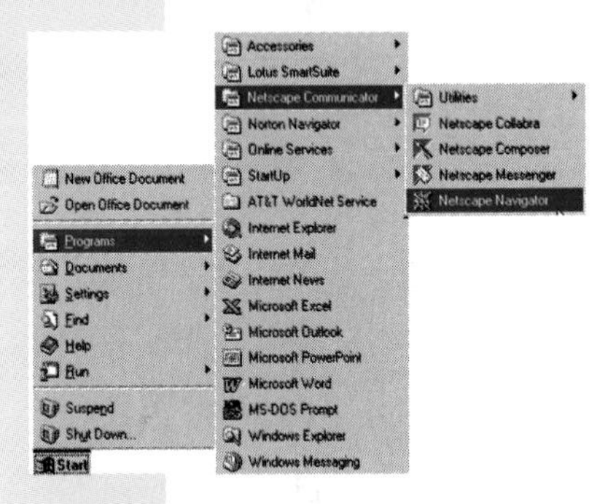

One Time Only: Create a User Profile ➤ *pp. 22-24*

1. When you see the New Profile Setup window, click Next.
2. Type your name.
3. Type your e-mail address.
4. Click Next.
5. Type a name for your profile—your first name, last name, or whatever you want to use to identify the profile easily.
6. Don't change the directory it's stored in.
7. Click Next.

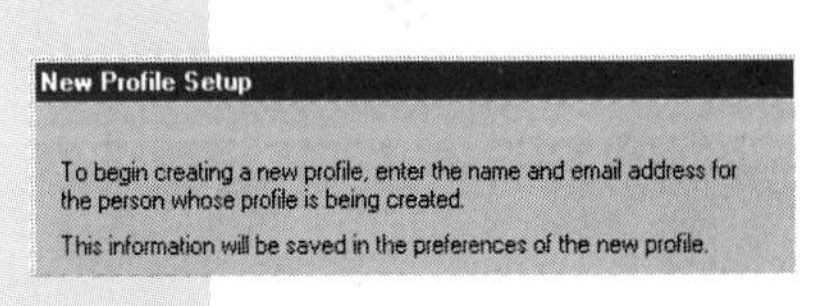

Click Smart ➤ *p. 26*

Once you see Netscape Communicator, use your mouse buttons like a pro:

- Only click once, never double-click a link.
- Right-click on anything to see a special pop-up menu.

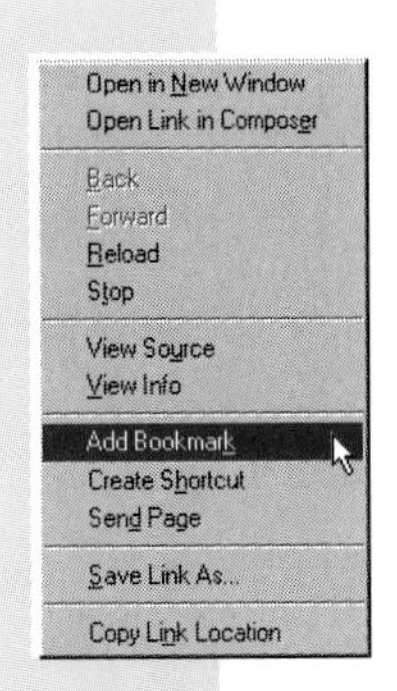

Worlds of information are available via the Internet and the Web. And if you're not on the Net, you're missing out. That's why you're reading this book.

How long will it be until you can start using Netscape Communicator 4.0? Our answer: almost immediately. Fortunately, we can take you through the heart of what you need to know in this book in a few evenings or lunch breaks.

We don't believe in spending a lot of time exploring philosophical underpinnings of, say, global communication. Instead, our aim in this book is to get you using Netscape Communicator 4.0 now. Of course, you can't use it if you don't have it yet. So we'll first show you how to choose the proper version of Netscape Communicator 4.0 and then how to download and install it.

Then, before you spend a lot of time customizing or fiddling with its features, we'll show you the basics of the program. You'll get a quick-click tour of the program and its features—not to mention how to use it. You'll send your first e-mail message and browse your first Web page. Then—if you're interested—we'll explore a little bit of Netscape's strategies and ideas and learn a little about Internet history.

Let's get started—you'll enjoy learning how each of the elements of Netscape Communicator 4.0 work, especially those you'll use most often: Netscape Navigator 4.0 (Web browser) and Netscape Messenger (e-mail and newsgroups).

Finding, Downloading, and Installing Netscape Communicator 4.0

If you don't already have Netscape Communicator 4.0, this section will show you how to choose the best version of the program for your needs, download the software, and install it. (If you already have Netscape Communicator 4.0 and have installed it, feel free to skip ahead to the next section.)

EXPERT ADVICE

If you're currently using a previous version of Netscape Navigator (which is now one of the components in Netscape Communicator 4.0), you'll want to upgrade.

CAUTION

We're going to assume you already have an Internet connection, either through a dial-up provider or via a corporate network.

Choosing the Best Version of Netscape Communicator 4.0 for You

As if life weren't complicated enough, Netscape has released Communicator in a passel of different versions (and we're not even counting the beta releases). Depending on how long you want to wait for the download, the features you need, and how you plan to use the software, you may choose any one of its different incarnations.

Netscape has also released Netscape Navigator 4.0 on its own, without the other stuff. If you've purchased Navigator from Netscape, we highly recommend you spend a little extra on Communicator (Standard)—the e-mail and discussion group capability, not to mention the Web page creation abilities, are well worth the price. If you got Netscape Navigator as part of a package on your PC or through your corporation, e-mail and other capabilities may be covered with other applications.

Table 1.1 lists the main versions of Netscape Communicator 4.0, the approximate file size of each, and the space required for installation, according to Netscape's estimations. Once you've tallied the system requirements against what you're willing to allocate to the program, read on to learn about what each version includes.

EXPERT ADVICE

The installed file size doesn't include the program's cache. (A cache is a portion of your hard drive Netscape Communicator 4.0 sets aside to store recently downloaded stuff.) You'll learn how to adjust the cache size later, but remember, you'll need to leave anywhere from 2 to 10MB of free space for the cache.

Version	Approximate size of downloaded file	Approximate space required for installation
Standard Edition ("Base Install")	8.3MB	16MB
Standard Edition plus plug-ins	11.6MB	20MB
Professional edition	18MB	40MB

Table 1.1 Versions of Netscape Communicator 4.0

If you purchased Netscape Communicator 4.0 in a retail store, you might have purchased it with some additional software as well. Netscape has created special editions of Netscape Communicator 4.0 for different needs. We won't cover all the extra stuff in this book—only the essentials and the stuff that everyone can use.

What's in the Standard Edition

Each version of Netscape Communicator 4.0 contains the core components found in the Standard Edition's "base install" package:

- Netscape Collabra, a module that allows a company or organization to set up internal "discussion groups." Collabra allows you to collaborate (thus the name) with others electronically, helping reduce meeting time and "phone tag." You'll find more information about Collabra in Chapter 7.

- Netscape Composer enables you to easily and quickly create Web pages and publish them to your Web site. You'll find more information about Composer in Chapters 9, 10, and 11.
- Netscape Messenger is a powerful Internet communications tool that enables you to send and receive e-mail as well as read and participate in Internet discussion groups (USENET). You'll learn more about Messenger in Chapters 6 and 7.
- Netscape Navigator 4.0, the most familiar and well-known component of Netscape Communicator 4.0, is the latest in its line of powerful Web browsers. Odds are you'll use this component most often. You can find more information about the newest version of Navigator in Chapter 3.
- Netscape Netcaster is the "tuner" you need to receive information that is "broadcast" across the Web. This new technology, often called *push technology*, automatically delivers information you've specified to your desktop. For more information on Netcaster, see Chapter 5.

If these components seem sufficient to you, or if you don't want to spend the time downloading more than 8.3MB, go ahead and choose this option—but it's probably worth the extra time to download the full Standard Edition. It's only a few megabytes larger and contains more options.

EXPERT ADVICE

A new feature in Netscape Communicator 4.0, called Smart Update, can automatically download program components as you need them, so if you choose the "base install" option, you'll still be okay.

If you choose the "full install" option, you'll be able to install these additional components:

- Netscape Conference helps you create virtual meetings where you, with a group, browse through documents, collaborate on a shared white board, share information, and hold real-time discussions.
- Bitstream font support allows you to view Web pages in which specific fonts have been "embedded"—where the pages' designers want you to be able to view their pages with the typefaces intact.
- Multimedia support modules perk up your Internet experience with support for sound, video, and 3-D environments (VRML).

For most people, the full version of Netscape Communicator 4.0 (Standard) is the best choice. However, you'll probably spend most of your time using the four main components of the Netscape Communicator 4.0 Internet suite, shown in Figure 1.1.

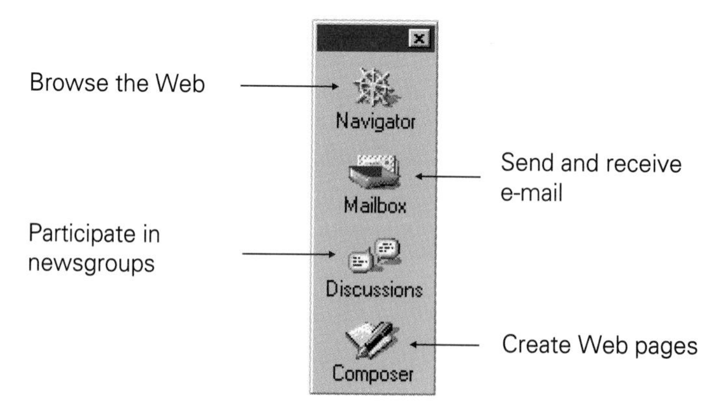

Figure 1.1 Netscape Communicator 4.0 sports a nifty toolbar that allows you to switch among the most frequently used modules

What's in Netscape Communicator 4.0 Professional?

You don't need the professional version unless you're part of a corporation or you need its advanced capabilities. You'll know if you need it because your corporate MIS department will have told you so, or you'll already be doing something fancy like terminal emulation. In any case, here's what the Professional edition adds to all the components listed above:

- Netscape Calendar helps you schedule tasks, meetings, and more across an intranet or the Internet.
- Netscape AutoAdmin helps system administrators install and maintain Netscape Communicator 4.0 installations.
- IBM Host on Demand provides access to IBM mainframes.

Go Get It: Downloading the Software

Once you've decided on the version of Netscape Communicator 4.0 you want, you can download it from Netscape's Web site.

You don't have to download the software if you don't want to. You might buy a copy from a software store off the shelf. Or you might get it on a CD in a magazine, or you might even have it already installed on your computer. Odds are you'll end up downloading it, if not now, then when Netscape releases a new, interim release (and there's no question about it—they will).

Here's the catch: you must already be connected to the Internet via a Web browser to download it (well, you could be using an FTP program, but you're probably not). However, in most cases, you'll be able to connect this way if you already have some kind of Internet connection.

From Netscape's home page (**http://home.netscape.com**), you can click on a download icon like the one shown here. Netscape occasionally changes how the downloading process works, but it's straightforward. Expect to complete the following steps.

GET THE LATEST NETSCAPE SOFTWARE

Tune Up to Communicator

Netscape Now!

Tune Up to Communicator

Get Any Netscape Software

For Subscribers Only

1. Specify the version of Netscape Communicator 4.0 you want to download.
2. Specify your computer's operating system (Windows 95, Macintosh, etc.).

3. Specify from where you want to download the software (you can pick from several sites—choose the one geographically closest to you).
4. Specify where you want to save the installation file (save it somewhere you won't lose it—like on your Windows desktop or in your My Documents folder).

CAUTION

Because Netscape Communicator 4.0 is such a big file, it's possible that you may run into problems with your dial-up Internet connection while you're downloading it. If the program you're using has a resume feature, you can just continue the download from where it left off when you disconnected. If you have repeated problems, talk a friend into giving you a copy on an Iomega Zip drive or other high-capacity disk format. And, finally, if your local connections are too busy and don't respond, try choosing one in a part of the world where it's the middle of the night!

EXPERT ADVICE

Don't forget: Netscape's licensing allows you to try the software, but if you're planning to use it, you should actually buy it—not just try it. You can purchase it online—you don't need to go to a store.

CAUTION

Even though you're anxious to get started with Netscape Communicator 4.0, actually downloading the software can take some time. Be warned: you may have to return to this book later, after it's downloaded.

Installing It

Once you've got the file, what do you do with it? Well, you've downloaded one big file—but it's not the file you'll use to *run* the program. Instead, you'll run this program to install Netscape Communicator to your computer.

To install the program, choose Run from the Windows Start menu. Click Browse and select the file from the folder in which you stored it. When you've found your downloaded file, click OK to close the Browse window. Then click OK to run the program.

When you've started the installation program, you'll see this window:

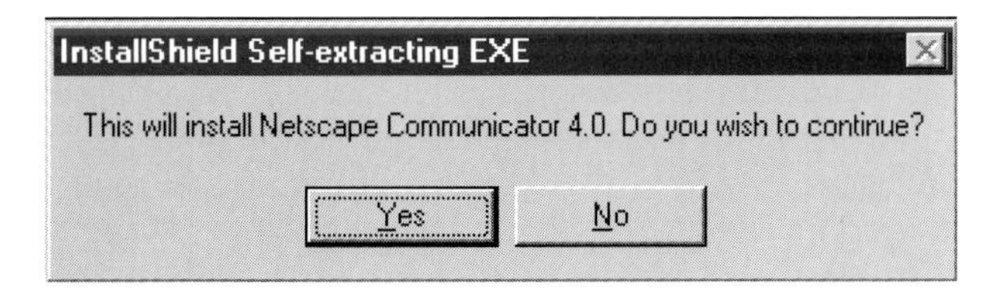

Choose Yes to get on with it. The installation program will first tell you it's extracting some files and then that it's preparing the Installation Wizard. *What* it's doing is irrelevant, really—all you need to do is wait until it asks you something. When it does, you'll see the screen shown in Figure 1.2.

Most setup programs ask you to close any other programs that you might have running—just in case the setup program tries to modify any files that your other programs have open. It's not essential that you close them, but it's a good

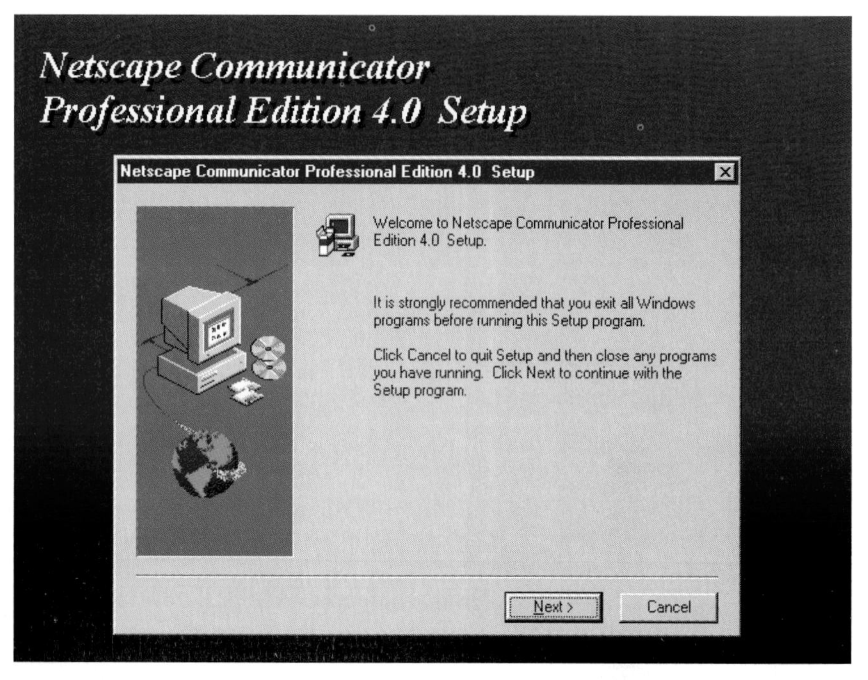

Figure 1.2 For best results, close all other programs when you begin Netscape Communicator's setup program

idea. You can press ALT-TAB to move among running programs and then exit each normally.

When you've closed any other programs, click Next. You'll then see Netscape's license agreement. Read the license agreement, make sure you agree, then choose Yes. Setup gets a little trickier now: you'll have to make some choices. Follow the Step by Step instructions to complete the installation.

EXPERT ADVICE

Even if you've closed all of the obvious applications, you may have some utilities like virus checkers running in the background. Don't worry too much about these. Go ahead with the installation and, if there's a problem, try eliminating these programs on your next installation attempt.

We recommend you go with the Typical installation. For the most part, you'll want the components Netscape suggests. And, unless you're drastically low on hard drive space—which means you're going to have a hard time running Netscape Communicator 4.0 anyway—there's no reason to begrudge it the room.

STEP BY STEP Off-the-Rack or Tailor-Made Installation

1 If you want to install Netscape Communicator 4.0's most commonly used components, choose Typical. If you choose this option, you won't have to choose which pieces of the program you want.

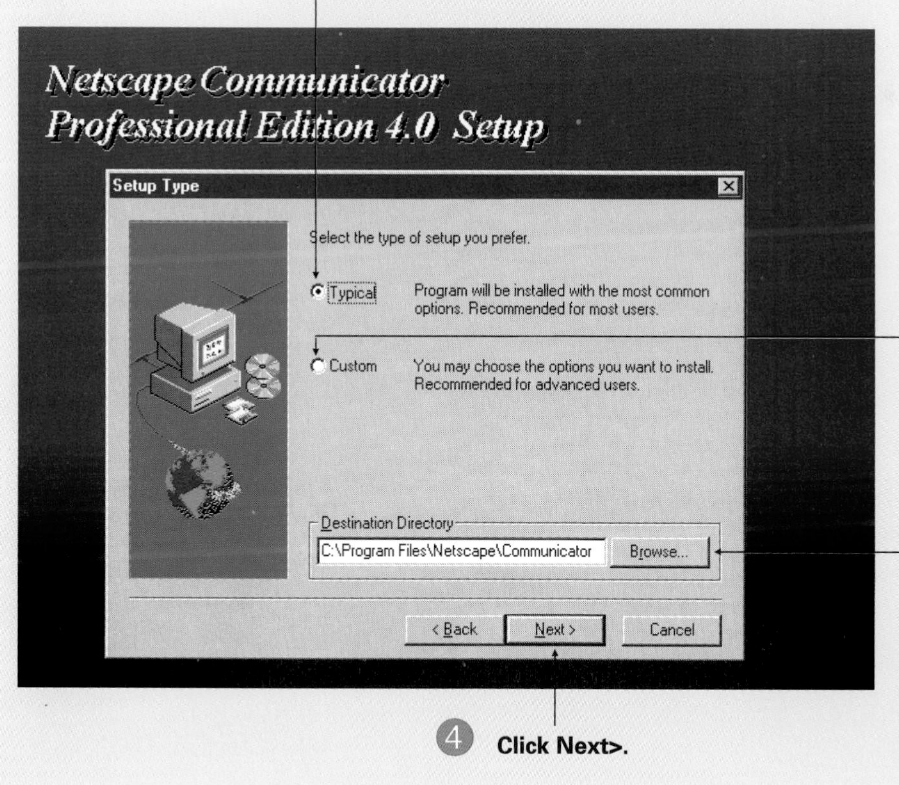

2 Or, if you want to choose exactly which program components you want to install, choose Custom.

3 Netscape Communicator 4.0 will be installed into the folder indicated in the Destination Directory box. If you want to change the folder, click Browse and select a different folder (or you can create a new one if you like).

4 Click Next>.

EXPERT ADVICE

If you already have an older version of Netscape Navigator on your computer, you can install Netscape Communicator 4.0 over it. But we recommend installing Netscape Communicator 4.0 to its default location and not erasing the previous version. Netscape Communicator 4.0 will copy settings from your previous installation—then when you're sure everything is working correctly, you can remove ("uninstall") the previous version.

If you do choose Custom install, however, you'll see a screen like this (this one's from Netscape Communicator 4.0's Professional edition):

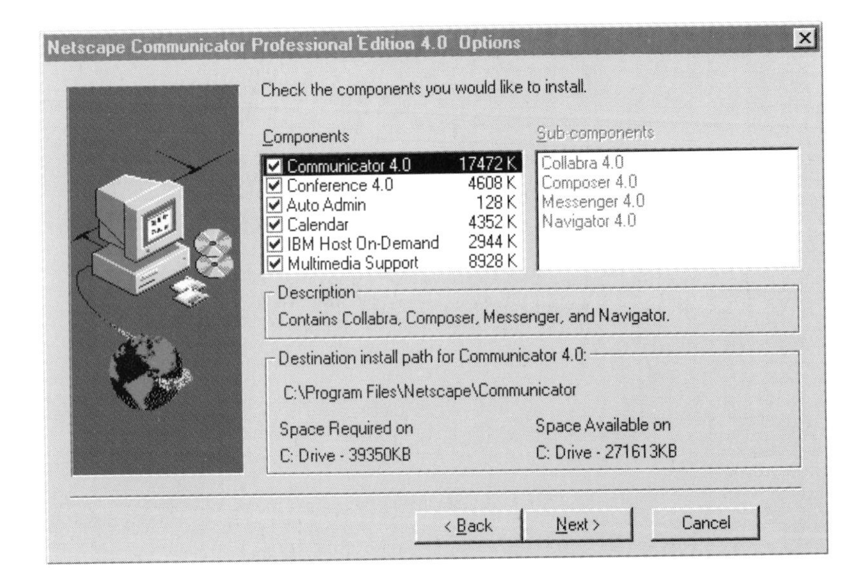

Check the components you want to keep (you'll definitely want to keep the Communicator component) and uncheck the others. For example, if you're planning to use Netscape Communicator 4.0 only for Web browsing and e-mail, you could select only the Communicator and Multimedia options. After you've chosen the components you want in your Custom installation, click Next.

Now, whether you've chosen a Typical or a Custom installation, the setup program will check for available disk space. If you already have a version of Netscape Navigator on your computer, the setup program will tell you this:

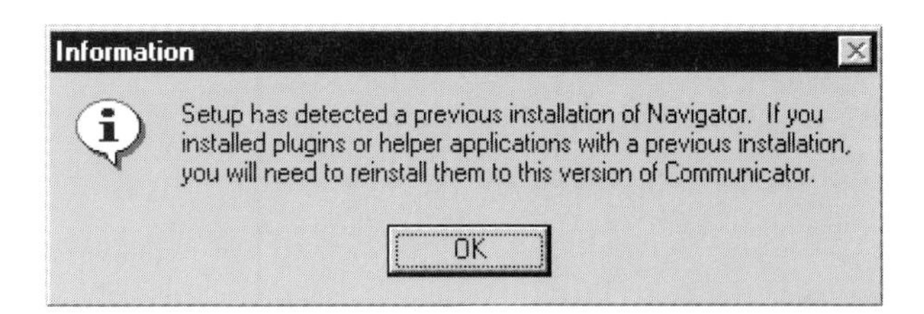

You won't want to stop the installation now to take care of these things, but you will probably want to pay attention to it later. In other words, if you had to install a special application to read certain types of files from the Web, you'll need to reinstall it to have it work with Netscape Communicator 4.0. For example, RealAudio—a "streaming" audio format that allows you to listen to real-time Web "broadcasts" of music, talk, etc.—requires you to install it anew when you upgrade to Communicator.

Follow the Step by Step instructions to tell Netscape Communicator 4.0 which types of files you want to open automatically using the program. Windows 95 can automatically open certain types of files using Netscape Communicator 4.0. These file types will be "associated" with Netscape Communicator 4.0. When you open them from Windows Explorer or from the desktop, they'll automatically be loaded into Communicator.

SHORTCUT

Each file type is checked by default—so if you want Communicator to open all the types, you can just click Next.

The installation program will ask you to choose a Windows 95 Start menu program group in which you want the Netscape Communicator 4.0 programs placed. We recommend you stick with Netscape Communicator—although you can pick anything you like: "Netscape,""Web apps," and so on.

STEP BY STEP Decide Which Types of Files You Want Netscape Communicator 4.0 to Open

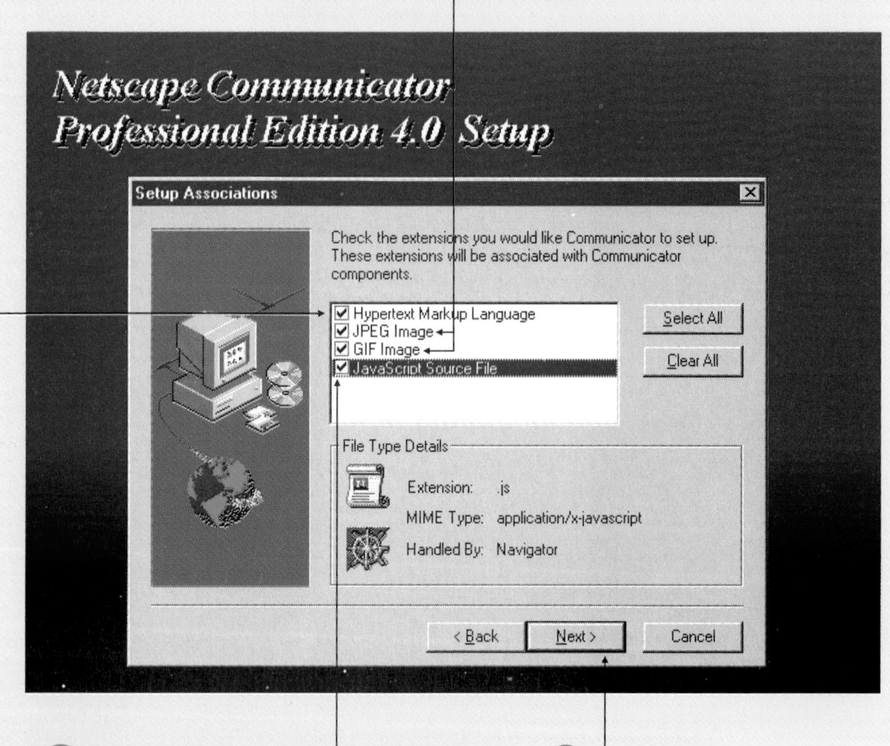

1. **If you want Communicator to be your default HTML file viewer (and editor), check here. It's generally a good idea to keep this one checked.**
2. **If you want Communicator to be your default viewer for GIF and JPEG files, check here. If you have another program that can edit—not just view—these graphics files, such as CorelDRAW or Windows 95's Imaging application, you may want to uncheck these.**
3. **If you want Communicator to open any JavaScript files you may come across, check here. You'll probably never encounter these in the wild, so just leave this one checked.**
4. **Click Next>.**

CAUTION

If you already have a Netscape Navigator group from a previous installation, you'll want to place the program icons for Communicator in a different programs group—it's easy to get confused when two icons have the same name.

Now, you've got one more chance to change any options. The setup program shows you a summary of what you've selected:

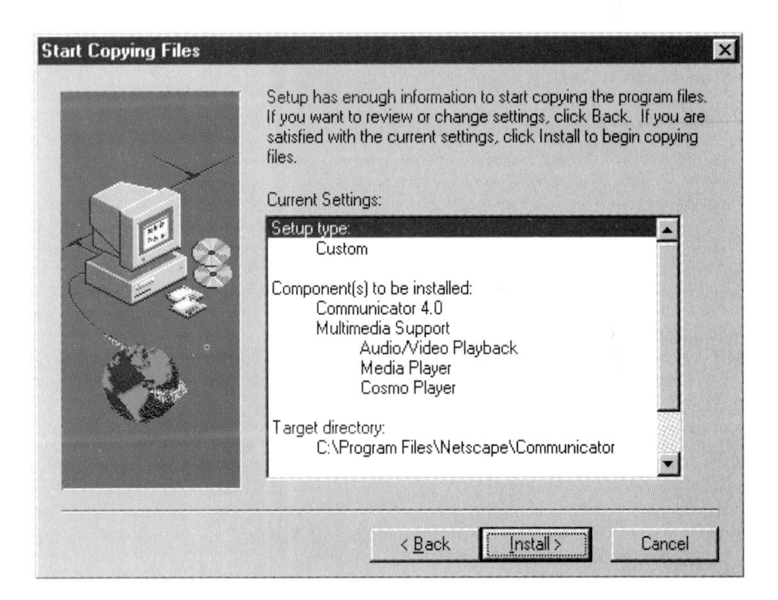

If you want to change anything, click Back. Otherwise, click Install to get on with the installation process (finally!).

The setup program will then install Netscape Communicator 4.0 to your computer. As it does, it'll keep you posted on its progress with a status update:

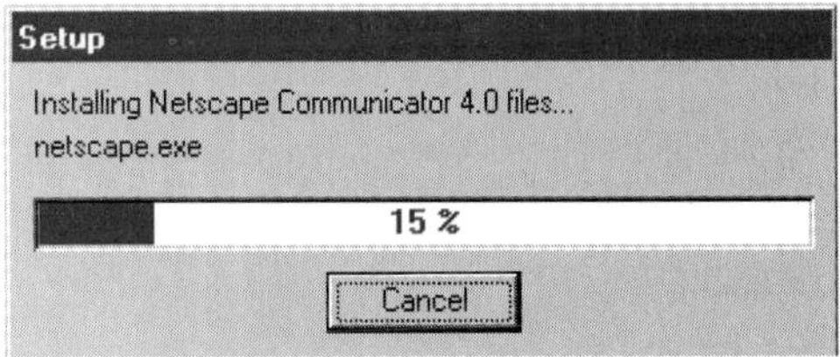

When it's done, you're asked if you want to read the "readme" file. These files often include useful information, so you might as well choose Yes. The setup program will load Windows Notepad to display the file. When you're through reading, choose File | Exit.

EXPERT ADVICE

Other computer programs include most of their release notes (information too new to appear in printed documentation) in the "readme" file. In this readme file, Netscape provides you with a Web address—a URL—that contains the most up-to-date information.

The installation program is almost finished! It will tell you it's finished even though there's still a little more to go:

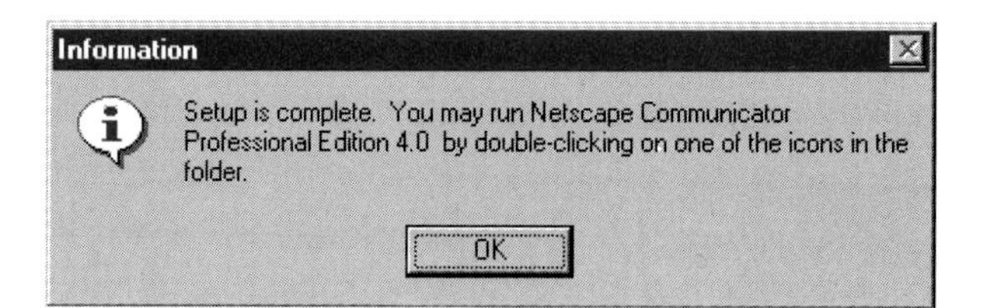

Click OK to see the final dialog box:

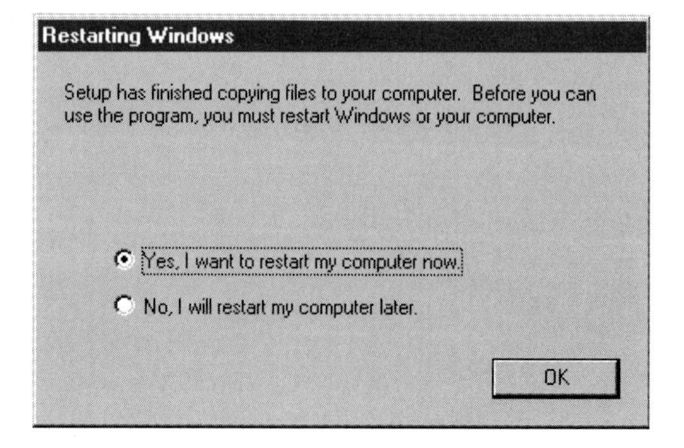

If you didn't listen to our advice and you have something else going on that you need to finish, choose "no" and restart your computer later. But, since you *did* listen to our advice, you might as well restart your computer *now* so we can get on with using Netscape Communicator 4.0. Click OK. Your computer will restart, and you'll be ready to use Netscape Communicator 4.0 (after a few minor adjustments, that is).

Your First Time: Running Netscape Communicator

You'll run Netscape Communicator the same way most of the time—by choosing it from the Start menu or double-clicking the icon on the desktop. (If you've set Netscape Communicator 4.0 as your "default" browser, it'll also automatically launch when you click a Web link in other applications like Microsoft Word or Corel WordPerfect.)

In the following sections, you'll first learn how to start the program. Then we'll take a quick look at creating a new "profile"—something you'll have to do only once. After that, we'll point you in the right direction to find out how to browse the Web, send e-mail, or do whatever you're trying to do.

Running the Program

You can launch Netscape Communicator in a couple of different ways:

- Click the Communicator desktop icon, *or*
- Choose Communicator from the Windows 95 Start menu.

If you double-click the desktop icon, Netscape Communicator 4.0 will open in Netscape Navigator—ready to browse the Web (you can change it if you'd rather have it open into newsgroups, e-mail, or even Composer).

If you'd rather choose which Netscape Communicator 4.0 module to start with, run it from the Windows 95 Start menu (Figure 1.3).

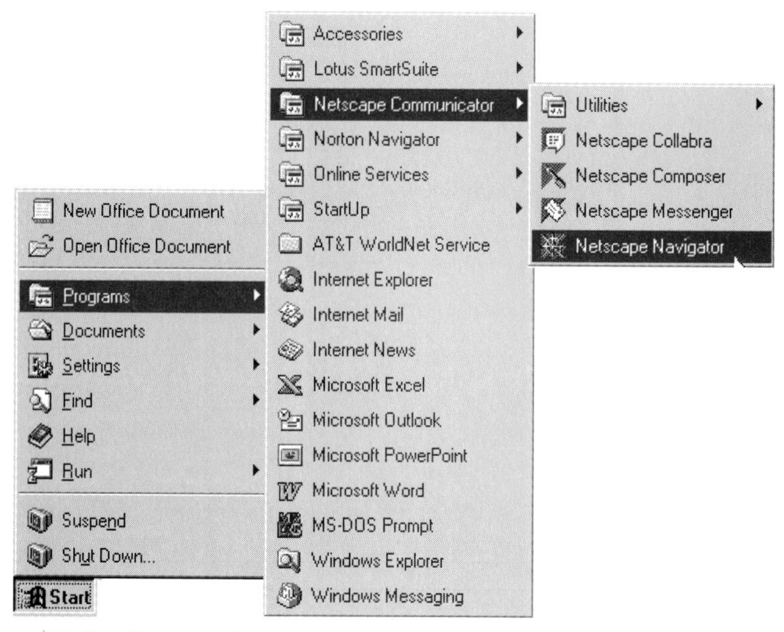

Figure 1.3 Run Netscape Communicator 4.0 from the Windows 95 Start menu to choose the module you want to use first—and access utilities, too

Choose Start | Programs | Netscape Communicator, and then choose the module you want:

- Netscape Collabra for Internet or intranet discussion groups and newsgroups (see Chapter 7)
- Netscape Composer to create Web pages (see Chapters 9–12)
- Netscape Messenger to send and receive e-mail (see Chapter 6)
- Netscape Navigator to browse and find things on the Web (see Chapters 3 and 4)
- Netscape Netcaster, if you've installed it, to automatically receive information from the Internet (see Chapter 5)

No matter which you choose, Netscape Communicator 4.0 will then launch.

Are You at Work or at Home?

If you're at work, you're probably attached to the Internet via your company network, so Netscape Communicator won't ask you if you want to connect to the Net.

But if you're at home, it's likely that you're using a modem to connect to the Net. Each time you launch it, Netscape Communicator 4.0 will ask you if you want to connect to the Internet by bringing up a Windows 95 dial-up networking window. Generally, you can just choose Connect.

EXPERT ADVICE

We've used Netscape Communicator 4.0 in a variety of ways—with an older, 28.8 kbps Hayes modem; with a low-priced, 36 kbps modem; and (most happily) with a state-of-the-art, 3Com/US Robotics x2 modem. Whichever you choose, make sure that your Internet provider (ISP) can support the type of modem you've chosen, especially if you have one of the newer, faster ones.

Netscape Communicator 4.0 comes with some neat features for working even when you're offline—the best one being Netscape Messenger, the e-mail component. See Chapter 6 for more details.

Exiting Netscape Communicator 4.0

"They've lost it," you're saying to yourself. "They're telling me how to exit, and I haven't even done anything yet." While we may have lost it, there's a good reason why we'll tell you how to exit the program now.

First, the easy stuff. To exit the program, choose File | Exit. Seems straightforward enough. But Netscape Communicator 4.0 can hang around even when you think it's gone. If you're running the e-mail module (Netscape Messenger) and the Web browser (Netscape Navigator), choosing File | Exit will close only one of those. You'll need to choose File | Exit in each module you're running to exit Netscape Communicator 4.0 entirely.

Status
Disconnect

You'll also want to make sure you've logged off from your Internet connection if you're using a modem. Under Windows 95, you can right-click the dial-up networking icon in the Windows 95 "tray" (the right side of the toolbar that the Start menu's in) and choose Disconnect from the pop-up menu.

If you are running an older version of Windows 95, you may not have the networking icon. You can also press ALT-TAB until your Dial-Up Networking window appears—then click Disconnect.

First Time Only: A New Profile

The first time—and only the first time—you run Netscape Communicator 4.0, you'll have to set up a new profile.

You can create multiple profiles to make it easy for more than one person to use the same copy of Netscape Communicator 4.0 on the same machine. See Chapter 2 for more details.

A profile sounds more ominous than it really is. It's unlike, say, a police profile: you won't have to report any unpaid parking tickets or divulge your shoe size. All you'll do is give Netscape Communicator 4.0 a few details—like the name you want to call the profile. Netscape Communicator 4.0 will then save all of *your* preferences in that profile—from connection preferences to bookmarks.

When you launch Netscape Communicator 4.0 for the first time, you'll see this window before you see any Web browser, e-mail, or other windows:

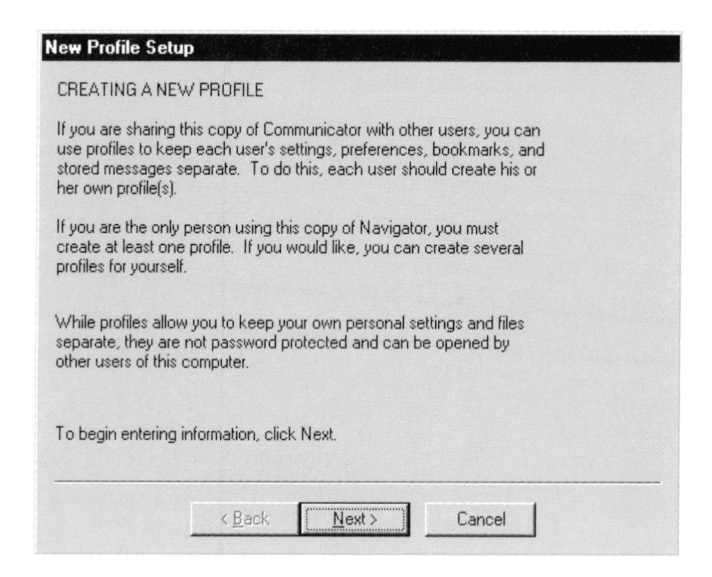

Click Next to get on to the (slightly) more interesting stuff.

To set up your first user profile,

1. Type your name in the Full name box.
2. Type your e-mail address in the Email address box.
3. Click Next>.

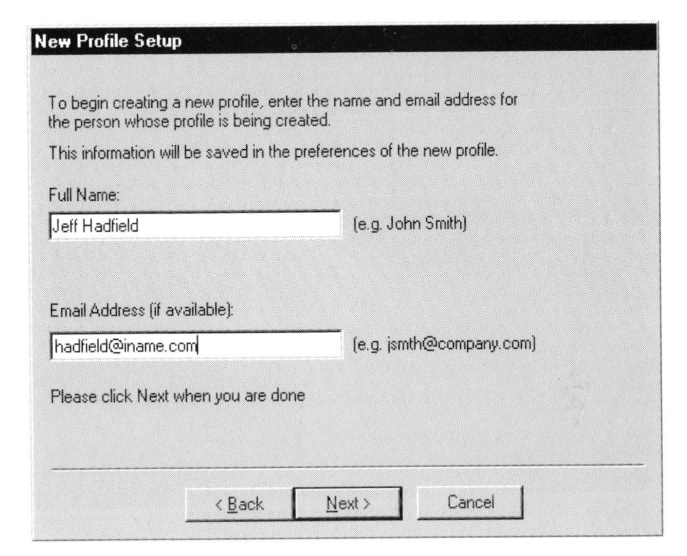

4. In the Profile name: box, type a name for your profile—your first name, last name, or whatever you want to use to identify it easily.
5. You can leave the other box alone—there's really no reason to change it from its default folder.
6. Click Next>.

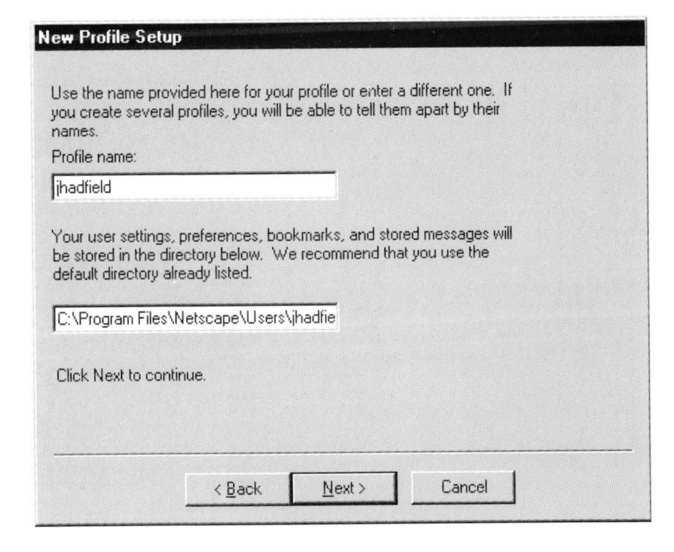

Is This for Me?

If you've *never* installed a previous version of Netscape Navigator or Netscape Communicator on this computer, skip ahead to "A Quick-Click Tour," below.

Watch out for one potentially confusing item, though: if you've had a previous version of Netscape Navigator or Netscape Communicator, your previous settings and bookmarks can be automatically transferred to Netscape Communicator 4.0. It's a good idea to do this and to accept the default suggestion, which moves your settings to the new directory (Figure 1.4). This way, previous versions (like Netscape Navigator 3) will share the same settings—and if you change them in one version, those changes will be reflected in the others. Make your choice and choose Finish.

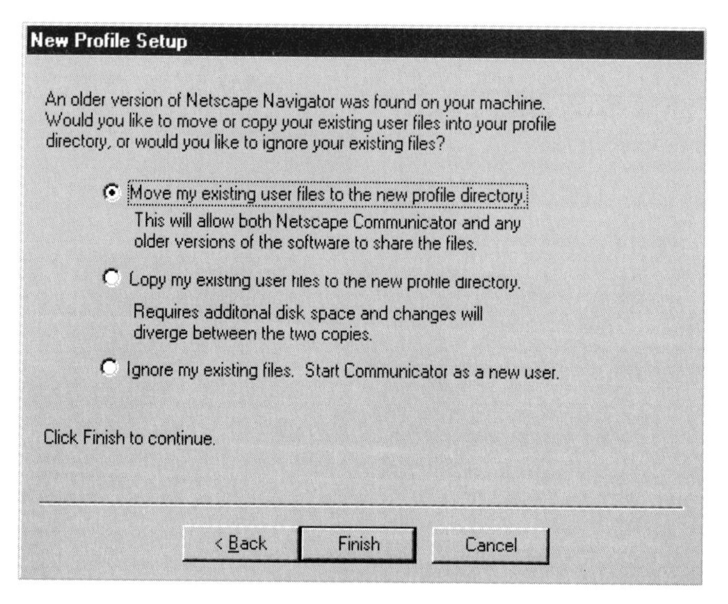

Figure 1.4 If you have a previous version of Netscape Navigator or Communicator, you'll see this dialog box, too. It's best to choose the top option (the default choice). Choose Finish to move on

A Quick-Click Tour

As you've already figured out, Netscape Communicator 4.0 is a collection of programs within a larger one. Launching one actually launches the others—they're just hidden from view. So now that you've launched Netscape Communicator 4.0 for the first time, what's there to see? Plenty.

It's time to get some things done: here's where to go next depending on what you want to do first:

- If you're the kind of person who wants things arranged "just so" before you start to work, read Chapter 2 first. It'll help you set up things just how you want them. If you're not that kind of person, come back to Chapter 2 after you've played around with the program for a while.
- If you're planning to use e-mail or newsgroups (Messenger or Collabra), you may want to read the relevant sections in Chapter 2, unless they're already set up for you (perhaps by your company's computer person).
- If you're ready to "surf the Web," "experience the Internet," and whatever other exciting phrase you can think of, you'll probably want to start with the Web browser: Netscape Navigator. Jump to Chapter 3.
- If you don't know where you want to go until you find it, go to Chapter 4: "Finding Things on the Web."
- If you want to send or receive some e-mail, go to Chapter 6.
- If you want to participate in discussions on your company's intranet or the Internet's newsgroups, go to Chapter 7.
- If you can't wait to build your own Web page, run over to Chapters 9–12 to learn about Netscape Composer.
- And, if you can't find what you're looking for in this list—we only listed the most common questions—check the TOC, flip through the index or, better yet, just launch Netscape Communicator 4.0 and see what you can do!

Click Smart

You know how to click the mouse. You can put those clicking skills to better effect in Netscape Communicator 4.0 with these quick tips to make you click-smart:

- You will *never* have to double-click on anything in a Web page. Click once if you want to activate a hyperlink to jump somewhere else.
- If you want to do something with anything you see in a Netscape Communicator 4.0 window—from a graphics image on a Web page to highlighted text in an e-mail message—try right-clicking (clicking the right-hand mouse button). In most cases, you'll see a special pop-up menu that will give you options custom-tailored to whatever you clicked on.

Is This for Me?

If you don't care about the whys or hows of Netscape Communicator and Netscape Navigator, just move along to another chapter. The following is background information that's nice to know but not essential.

Who, or What, Is Netscape?

It's easy to get confused: most people say "Netscape" when they really mean "Netscape Navigator," the company's best-known product. Netscape, the company, provides software for Web servers and software for people who want to use the Internet. This software for Web users includes Netscape Communicator 4.0—which, in turn, includes Netscape Navigator.

Why is Netscape Communicator free or cheap? How can they give it away free, and why do they? Well, they really don't give it away free, unless you're bold enough to live on the cutting edge and continue to use beta versions. If you download the product, you should pay for it after the trial period is over—or just use your credit card right when you download it from Netscape's Web site.

But Netscape's money is made on the server side (also through other ways), so it's a good deal for them. They also make money from companies who license the software for their employees to use.

Netscape vs. Microsoft

Unless you've been hiding under a very large rock and avoiding any sort of technology or business news, you've heard about the great battle between Netscape and Microsoft for the title of Most Popular Web Browser. (Depending on the survey data you read, one or the other might be more popular, but most agree that Netscape's still holding a majority.)

It's more than a popularity contest, though: each company is trying to compete with the other by having better features, niftier interfaces, and more industry partners committed to use their browser as the "preferred" interface for their Web sites.

What does all this mean to you? Well, on the positive side, it means that neither company can rest on its laurels for long. Netscape beat Microsoft to market with Netscape Communicator 4.0—but Microsoft's Internet Explorer 4.0, because of its delay, will have some features Netscape Communicator 4.0 doesn't. Of course, Netscape will soon come out with an updated version of Netscape Communicator 4.0 (at this writing, you can probably expect to see Netscape Communicator 4.2 as the immediate response to Internet Explorer 4). This constant one-upmanship makes for an exciting market, but you'll want to keep your eye out for the latest versions.

There are almost as many disadvantages as advantages, at least for now. When the products change frequently, it means you have to keep up on the latest news to make sure you have the latest and greatest. Of course, if you're so far on the cutting edge that you're using beta software, you're basically testing stuff that's not ready for prime time.

You'll also find that Netscape and Microsoft haven't quite yet agreed on a consistent way of presenting or preparing information for the Web. HTML—hypertext markup language—at its most basic can be read by either browser. But both companies have proprietary extensions to HTML that the other product can't read—at least, not without some help. Our take: expect this to settle down a bit as the user community and technology dictate that the two sworn enemies compromise on a consistent set of standards.

For now, while Microsoft attempts to integrate its Internet abilities into the operating system (Windows 98), Netscape Communicator already offers a complete suite of Internet tools in one program—without locking you into one set of tools. In short, you'll be able to view just about anything you'd ever want to see using Netscape Communicator 4.0. You don't need both browsers on your system.

EXPERT ADVICE

Don't install IE and Communicator on the same system unless you feel that you're experienced enough to deal with any minor issues that might arise. Microsoft—perhaps in preparation for Windows 98—encourages you strongly to make IE the default Web browser, the default HTML file viewer, etc., and it'll ask you these things tenaciously.

You now know how to install Netscape Communicator 4.0 and run it for the first time. You also have a good idea of what Netscape Communicator 4.0 can do for you. Now, jump ahead to the chapter that most interests you, or read on to Chapter 2 to set a few additional preferences.

CHAPTER

2

Stuff You'll Need to Do Only Once

INCLUDES

- Setting your Web browser preferences
- Changing default colors and fonts
- Configuring your e-mail account
- Configuring your newsgroups
- Customizing Netscape Communicator 4.0's interface
- Working online and offline

Set up Browser Preferences ➤ *pp. 33-40*

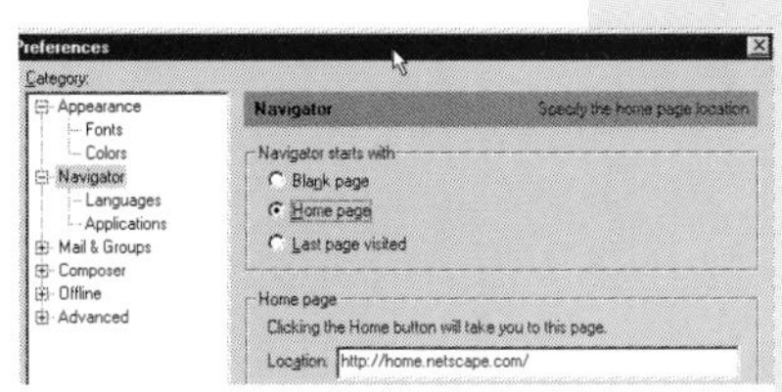

1. Choose Edit | Preferences.
2. Choose the type of preferences you want to adjust from the Category menu.
3. Adjust the settings as you like.
4. Choose OK.

Make the Toolbar Display Only Text ➤ *pp. 36-38*

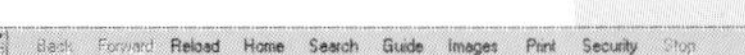

1. Choose Edit | Preferences.
2. Choose the Appearance category.
3. In the Show toolbar as group, check Text only.
4. Choose OK.

Choose Default Web Page Fonts ➤ *pp. 38-40*

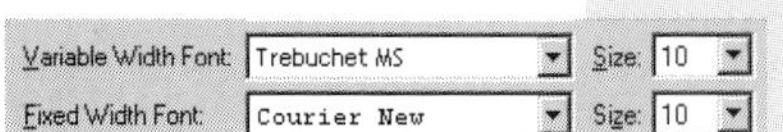

1. Choose Edit | Preferences.
2. Choose Appearance | Fonts.
3. Choose the variable-width font you want to use for regular text.
4. Choose the fixed-width font you want to use for monospaced text.
5. Choose the radio button indicating how you want to handle pages that provide their own fonts.
6. Choose OK.

Set up Your E-mail Account ➤ *pp. 40-42*

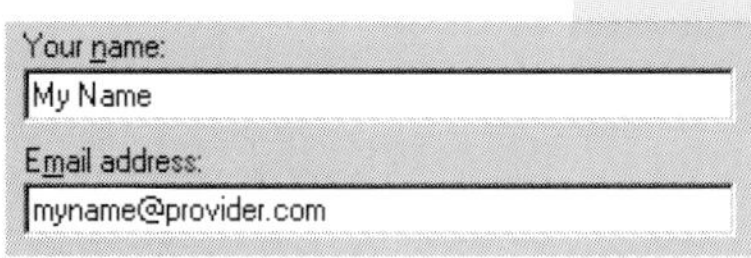

1. Choose Edit | Preferences.
2. Choose the Mail & Groups | Identity category.
3. Type your first and last name or whatever your full name consists of.
4. Type your e-mail address.
5. If you want people to reply to a different e-mail address, type it.
6. Type your company or organization name (if you want).
7. If you would like a block of information—a signature—to automatically appear at the bottom of your messages, enter the filename of the text or HTML file.

8. Check the box to automatically attach your Netscape Communicator 4.0 "address card" if you want.
9. Choose the Mail & Groups | Mail Server category.
10. Type your mail user name.
11. Type the name of the server that distributes outgoing mail for you, then the name of the server on which mail coming *in* to you is kept.
12. Select a mail server type.
13. Choose OK.

Set up Your Net News ➤ *pp. 42-43*

1. Choose Edit | Preferences.
2. Choose the Mail & Groups | Groups Server category.
3. Type the name of the Discussion groups (news) server you've been given.
4. Choose OK.

Mail & Groups
Identity
Messages
Mail Server
Groups Server
Directory

Set up User Profiles ➤ *pp. 43-46*

1. Exit Netscape Communicator.
2. From the Windows 95 Start menu, choose Programs | Netscape Communicator | Utilities | User Profile Manager.
3. To add a User Profile, click New.
4. To change the name of a User Profile, click Rename.
5. To remove a profile, click Delete.

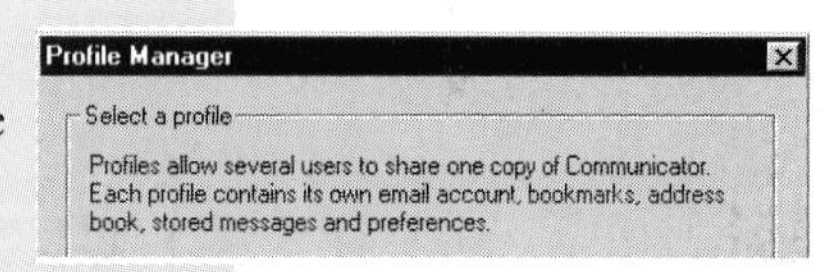

Work Online or Offline ➤ *pp. 46-48*

1. Choose Edit | Preferences, then the Offline category.
2. If you're always connected to the Net, choose Online Work Mode.
3. If you're going to connect when needed, choose Offline Work Mode or Ask Me.
4. Choose OK.

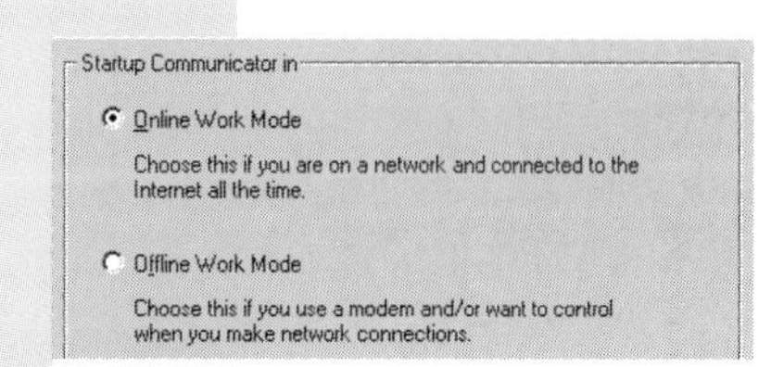

This chapter teaches you stuff you will only have to mess with once. Some of it helps Netscape Communicator 4.0 talk to your network or Internet service provider. Other information helps you arrange things just how you like them.

Just like when you get a new car, you need to adjust things in Netscape Communicator so they're how you like them. The basics in a new car, such as gas, brakes, and steering wheel, apply no matter what car you drive, but you'll adjust the seat back, seat height, and steering wheel height. You'll turn the radio to your favorite station and program all the presets. You'll adjust the mirrors and fiddle with the climate controls. You may even change the look by adding a bumper sticker, personalized license plate, or license plate frame.

Is This for Me?

If you're working in a company where your IS department sets everything up for you—or if you're just not comfortable changing settings in your software—most of this chapter won't apply to you. If you need to set up your e-mail and newsreader connections, go ahead and skip to that section.

You'll learn how to set up the basics and to custom-tailor Netscape Communicator 4.0 to your liking. Later, you'll learn how to operate each feature of the product, but for now, let's concentrate on the stuff that must come first.

Make Your Browser Work the Way You Like It

You can tailor Netscape Navigator in a bunch of ways, but you'll really only need to worry about a few of the settings. You can adjust all of Netscape's preferences from the same dialog box (see Figure 2.1). To get there, just choose Edit | Preferences.

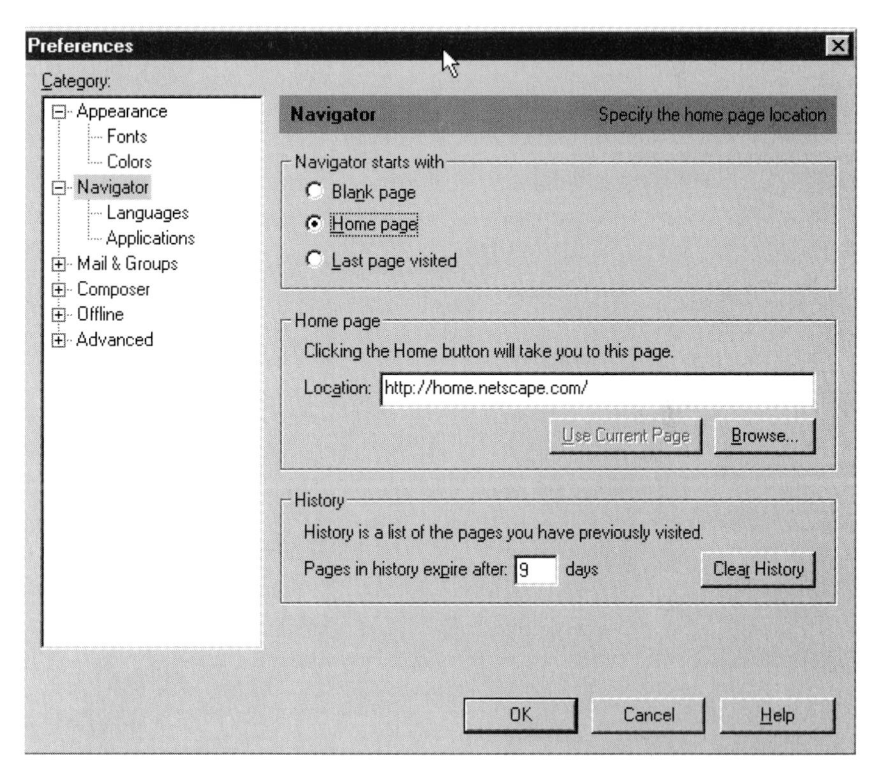

Figure 2.1 Tailor Netscape Navigator to your liking using the Preferences dialog box

Let's look at a few of the most important—and a few of the most fun—settings to change.

Tell Netscape Where to Start

A few of these settings—the ones you see in Figure 2.1—help you adjust how Netscape Navigator starts. First, if you're using a modem to dial the Internet, you might want to open onto just a blank page when Netscape Navigator loads—an empty window that doesn't require you to connect to the Net. Of course, if you choose Last page visited, Netscape Navigator will try to load up your default dial-up connection.

You might want to change your home page to a different one if you choose Home page (or you might just feel like changing it anyway). Of course, Netscape likes you to use theirs, so that's what comes as the default. It's a nice way to keep up on what's happening with Netscape, but, if you're like us, there are other pages you might use more frequently—like your company's intranet.

EXPERT ADVICE

If you prefer, you can use an HTML file from your local hard drive as your home page. These load quickly and can provide a customized launching pad for the sites you view often. Just use file:// for the protocol and then enter a path and filename for the HTML document you wish to use. You can even use your bookmarks file, which is an HTML file, as your start page.

To change your default home page, just type the URL in the Location text box (you might want to use copy and paste to guard against typos). If you're playing with these settings while you're on the Web, you can just surf to the page you want to use as the home page and click the Use Current Page button.

EXPERT ADVICE

*It's a good idea to type the full URL in dialog boxes like this one. In other words, be sure to type the full http:// prefix—**http://www.wpwin.com** instead of just **www.wpwin.com**.*

As long as we're on this page, a quick explanation of the final setting: You can adjust the length of time you keep a page history. A page history—basically just a list of pages you've visited (Figure 2.2)—can go back any number of days you specify. It's probably a good idea to keep it around for a week or so, though. The paranoid can clear the history file by clicking Clear History.

History

File Edit View Communicator Help

Title	Location	First Visited	Last Visited	Expiration	Visit Count
Bay Area Rapid Tran...	http://www.bart.org/	5 days ago	5 days ago	8/5/1997 9:19 PM	1
Getting Around the B...	http://www.metrodynamic...	5 days ago	5 days ago	8/5/1997 9:17 PM	1
Welcome to the Bay ...	http://www.metrodynamic...	5 days ago	5 days ago	8/5/1997 9:16 PM	2
Yahoo! Search Results	http://search.yahoo.com/...	5 days ago	5 days ago	8/5/1997 9:15 PM	1
Yahoo!	http://www.yahoo.com/	5 days ago	5 days ago	8/5/1997 9:14 PM	1
City.Net Bay Area, Ca...	http://city.net/countries/u...	5 days ago	5 days ago	8/5/1997 9:13 PM	1
Excite Search Result...	http://www.excite.com/se...	5 days ago	5 days ago	8/5/1997 9:12 PM	1
Excite	http://www.excite.com/	5 days ago	5 days ago	8/5/1997 9:12 PM	1
Give us your ideas!	file:///C|/FILES/web book...	5 days ago	5 days ago	8/5/1997 9:07 PM	1
Internet Book Series	http://www.osborne.com/i...	5 days ago	5 days ago	8/5/1997 9:04 PM	2
The World Wide We...	http://www.osborne.com/i...	5 days ago	5 days ago	8/5/1997 9:03 PM	1
Welcome to Osborne...	http://www.osborne.com/...	5 days ago	5 days ago	8/5/1997 9:02 PM	3
Osborne/McGraw-hill:...	http://www.osborne.com/...	5 days ago	5 days ago	8/5/1997 9:02 PM	1
	http://www.osborne.com/l...	5 days ago	5 days ago	8/5/1997 9:02 PM	3
Osborne/McGraw-Hill...	http://www.osborne.com/...	5 days ago	5 days ago	8/5/1997 9:02 PM	1
Osborne/McGraw-Hill	http://www.osborne.com/...	5 days ago	5 days ago	8/5/1997 9:01 PM	1

Netscape

Figure 2.2 Use the History window to jump quickly to pages you've visited recently

EXPERT ADVICE

To access your history file in Netscape Navigator, choose Communicator | History. You can sort the file by date, URL, or any of the listed columns. You'll also find that History is particularly useful for creating bookmarks after the fact—when you wish you'd bookmarked a page originally.

How Much Cache Can You Afford?

Most people don't realize that Web browsers like Netscape Navigator use more disk space than just the program files. In fact, Netscape Navigator uses a cache that can take up megabytes of disk space. But what is this cache used for?

Mainly, it saves Netscape Communicator 4.0 (and you) a little time when returning to pages you've visited recently. If the page hasn't changed in the meantime, Navigator can open the cached (local) copy of the document that originally came from a remote server.

DEFINITION

Cache: *An area on your hard drive Netscape Communicator 4.0 sets aside for temporary files and elements of recently visited Web pages. There's also a memory cache that stores stuff in your RAM and takes more memory than just the program itself.*

This is an advanced setting, but it's pretty darn essential, so you'll need to know a little bit about it. Get yourself to the proper place in the Preferences dialog box by choosing Advanced | Cache from the menu tree on the left side of the dialog box (Figure 2.3).

You can specify the amount of both physical memory and the amount of disk space Netscape Navigator will use. If you're using a PC with less memory—say, only 8MB—you may want to reduce the Memory Cache to 512K if you're planning to run other applications at the same time. If you're using some obscenely fast PC with a whole bunch of RAM—like 32MB or more—go ahead and increase it if you like.

By the same token, if you have a huge hard drive and plenty of free space on it, you can increase the size of the disk cache. The disk cache helps the pages that you view frequently load faster by storing page elements on your machine so you don't have to download them every time.

Fiddlin' with the Toolbars

While you're customizing Netscape Communicator 4.0 to work smoothly on your machine, here's another quick tip. If you're running it on a smaller display such as a laptop, you may want to save screen real estate by changing your toolbar to display only text—not both pictures and text. To do so, choose Appearance

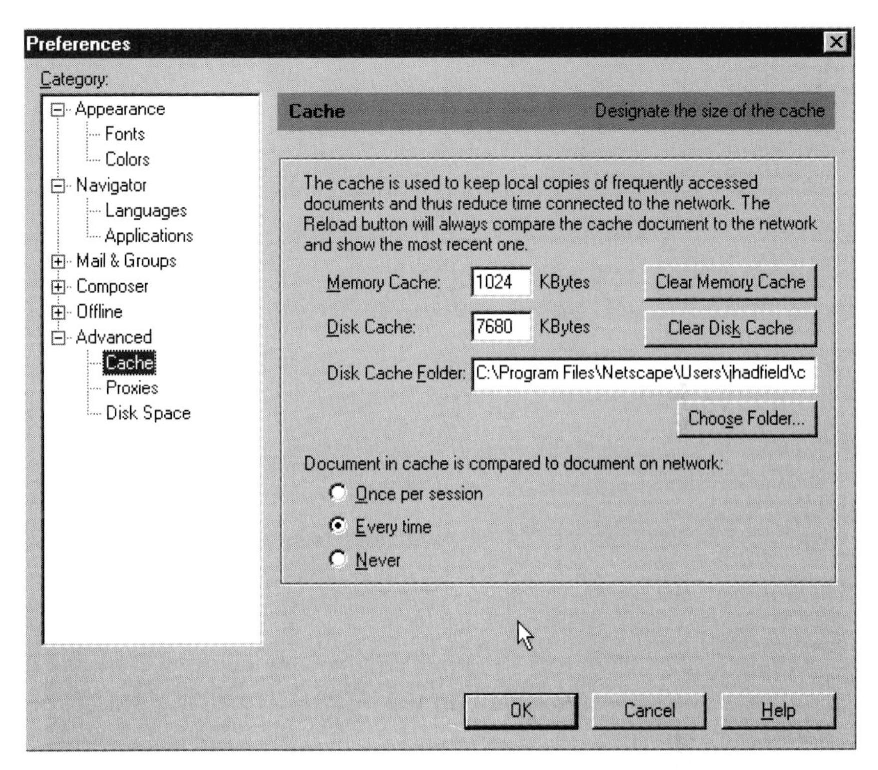

Figure 2.3 Set your memory and disk cache to best fit your PC's configuration

from the menu on the left side of the Preferences dialog box, then choose the appropriate check box in the Show toolbar as group (Figure 2.4).

Remember, you can always temporarily reduce the toolbars to the little tabs any time you want more screen real estate.

EXPERT ADVICE

The bottom bar is your Personal Toolbar. You can add any bookmark—a shortcut to any Web site—to this toolbar. Just navigate to the page you want to add, then click Bookmarks | File Bookmark | Personal Toolbar Folder | Personal Toolbar Folder. Edit the folder as you would your bookmark list (see Chapter 3 for more details).

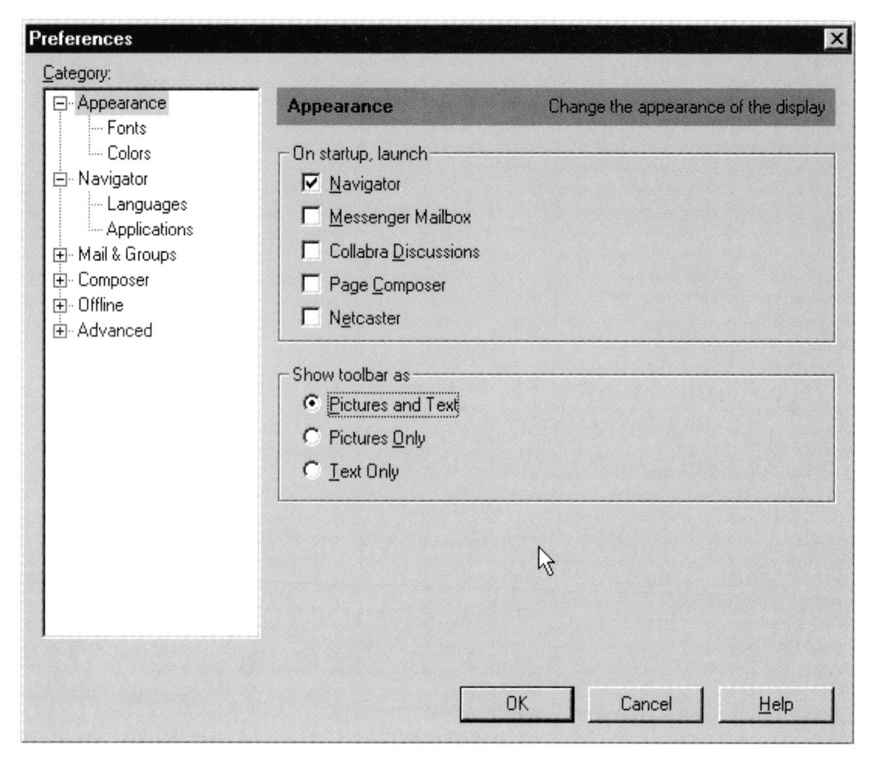

Figure 2.4 Customize your toolbar's appearance

On a laptop, for example, you'll probably want to choose just text, like this:

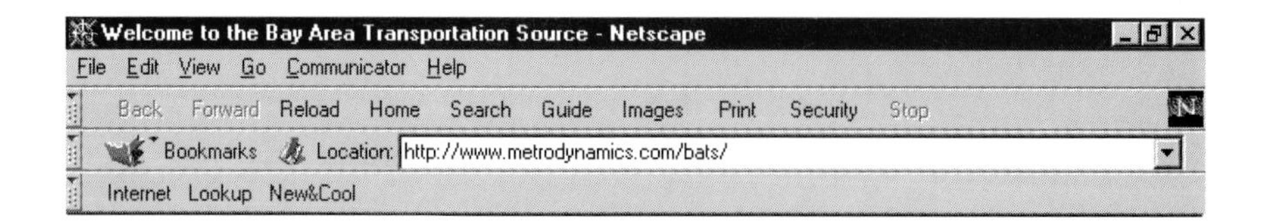

Choosing Fonts and Colors That Are Easy on Your Eyes

Usually, Web pages don't have any font specified; it's your browser, Netscape Navigator, that's assigning a font by default. The default is Times Roman. It's nice, but plain. You may prefer a sans serif typeface like Lucida Sans (some studies suggest sans serif fonts are easier to read on a computer screen). Christian suggests Microsoft's new (free) Verdana. Jeff's favorite type (which you'll see throughout the book) is Trebuchet, a typeface that's also available free from Microsoft. He's also used some neat typefaces by Adobe and Bitstream. Experiment—you can always change the setting!

When you change the setting, you can specify two types of text: variable-width and fixed-width. Variable-width text is like what you're reading: the spaces between the letters vary depending on the shape of the letter.

`Fixed width text, on the other hand, is like the text in this sentence: the space between letters is identical, regardless of letter shape.`

STEP BY STEP Choose Default Web Page Fonts

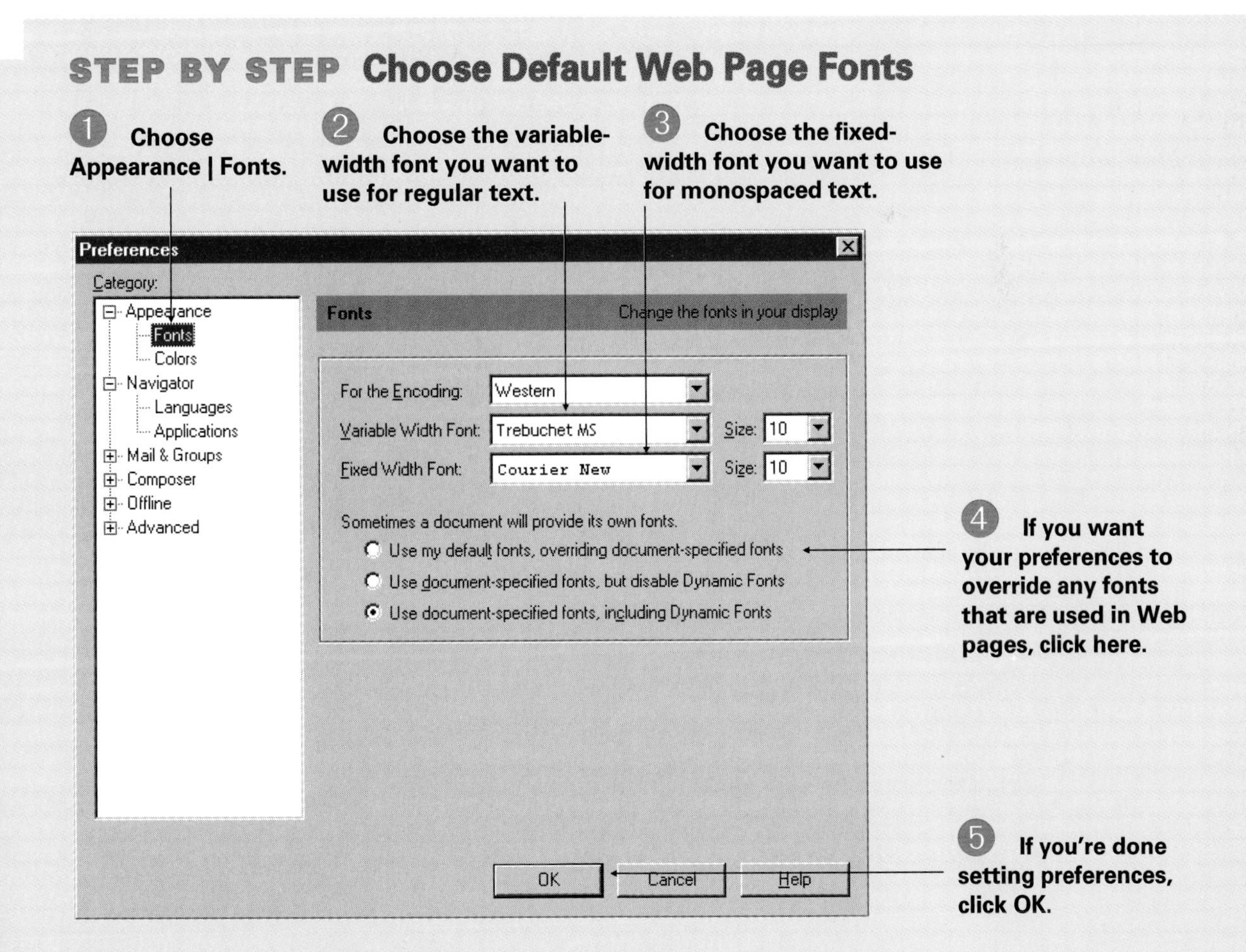

CAUTION

Unless you really feel strongly about the fonts you choose, you should leave the bottom settings alone. Choosing to override fonts that are used in Web pages you view can make those pages appear, well, downright goofy.

You might want to play with some of the other settings available in the Preferences dialog box, including changing the color of hyperlinks, etc. For example, you may want to increase the base font size to 12 or 14 if you're working on a small screen or have weak eyes.

EXPERT ADVICE

You can also adjust the base font size of your Web pages while you're viewing them. Just press CTRL-[or CTRL-]*, or choose View | Increase font or View | Decrease font.*

DEFINITION

Dynamic fonts: *Those specified in the original Web page. They're transmitted to your computer to display pages exactly as designed—but they can significantly increase the time needed to download the page.*

Enter Your Identity for Mail and News

If you're using a modem to connect to the Internet, your access provider gave you some information about your e-mail and news servers. Go find it before you try this section. (If you're on a LAN—that is, you have a direct connection—check with your system administrator.)

SHORTCUT

If you've set up any other e-mail program on this computer or any other nearby computer, you can probably copy the pertinent information.

The easiest way to set up your e-mail and news accounts is to use the Wizard that walks you through it the first time you run Netscape Communicator. This Wizard holds your hand through entering the information we'll enter in a different place: the Preferences dialog box (remember, once we finish all these settings, you don't have to come back here again).

Choose Edit | Preferences. Then, from the Category list on the left side of the window, choose Mail & Groups | Identity. Remember, your Internet service provider probably gave you many of these settings.

STEP BY STEP Setting Your "Identity" for E-mail and Newsgroup Messages

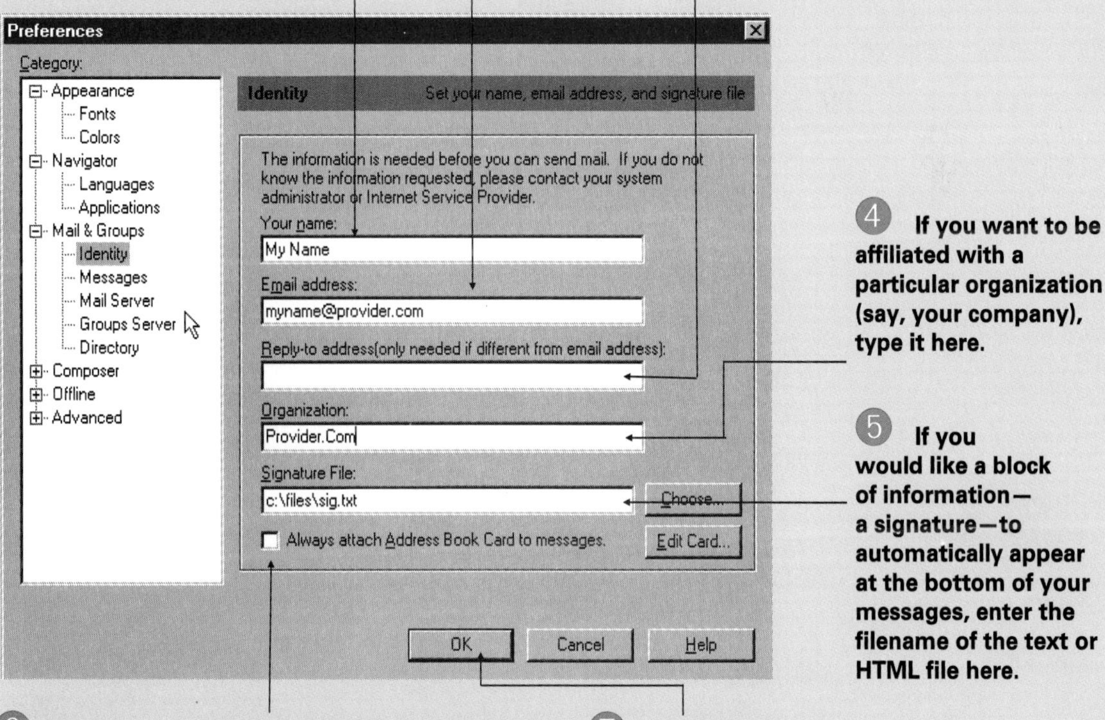

DEFINITION

Signature file: *Usually includes information such as your name, phone number, organization, e-mail address, etc. Your signature file will probably depend on what information you want to share with everyone—business contact information is generally more widely shared than home or personal information.*

EXPERT ADVICE

To create a signature file, you can create the text you want to use in Netscape Composer or even in Windows Notepad. Be sure to save the file you create in either plain text or HTML format. Confine your signature to a handful of lines, generally no more than six or seven, to stay within the bounds of proper netiquette.

Your E-Mail Server(s)

Your e-mail doesn't just find its way across the Internet to your computer as if it were carried by homing pigeons. Instead, it's stored on a server—a computer that "serves" other computers with information—and then distributed to individual users. In order to get your e-mail, you'll need to enter a few bits of data into the Preferences dialog box, shown in Figure 2.5. To get there, choose the Mail & Groups | Mail Server category.

Type the three key bits of information you've gotten from your Internet service provider: your mail user name, the name of the server on which mail coming *in* to you is kept, and the name of the server that distributes outgoing mail for you. Don't worry about the other settings: most ISPs use POP3 mail protocol, and those that don't (like MSN) will be doing so soon.

Your News Server

If you're planning to read USENET newsgroups (more about those in Chapter 7), you'll need to set up a few things first. Again, your Internet service provider (ISP) should have given you all this info already, so it shouldn't be too hard.

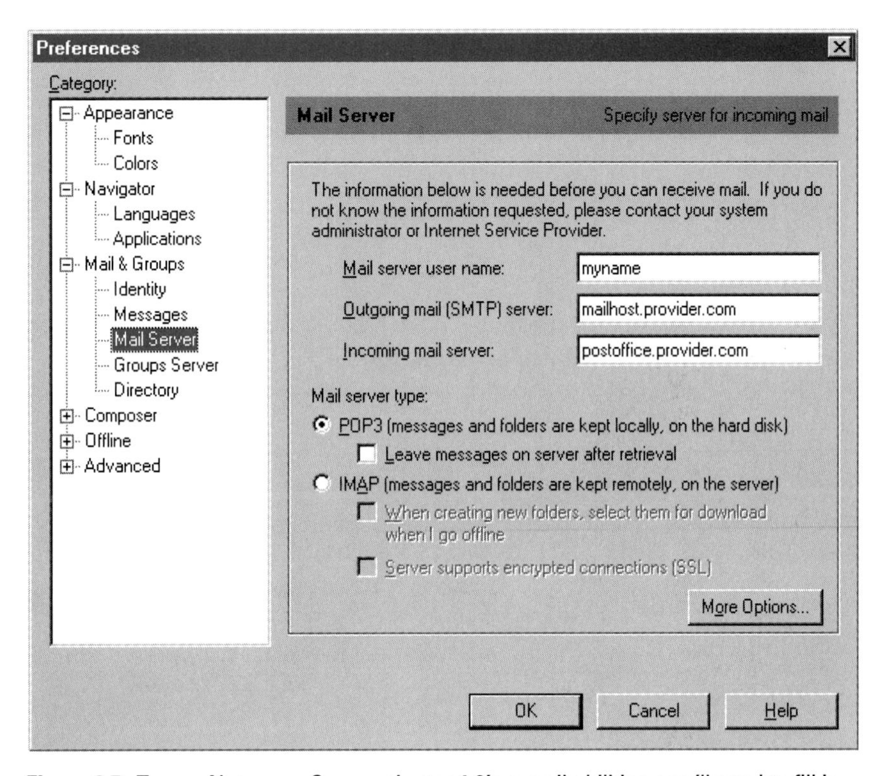

Figure 2.5 To use Netscape Communicator 4.0's e-mail abilities, you'll need to fill in these key bits of information

If you're not already in the Preferences dialog box, choose Edit | Preferences, then choose the Mail & Groups | Groups Server category to get to the dialog box shown in Figure 2.6. (No, we don't know why they call it "groups" instead of "news.")

As a busy person, you'll really only need to worry about one setting in here. Type the name of the news ("groups") server you've been given.

You and Who Else?

If you did what we asked you to do when you installed Netscape Communicator 4.0, you just said okay and got through it as fast as you could. One of the things we didn't worry about then was what Netscape calls "user profiles."

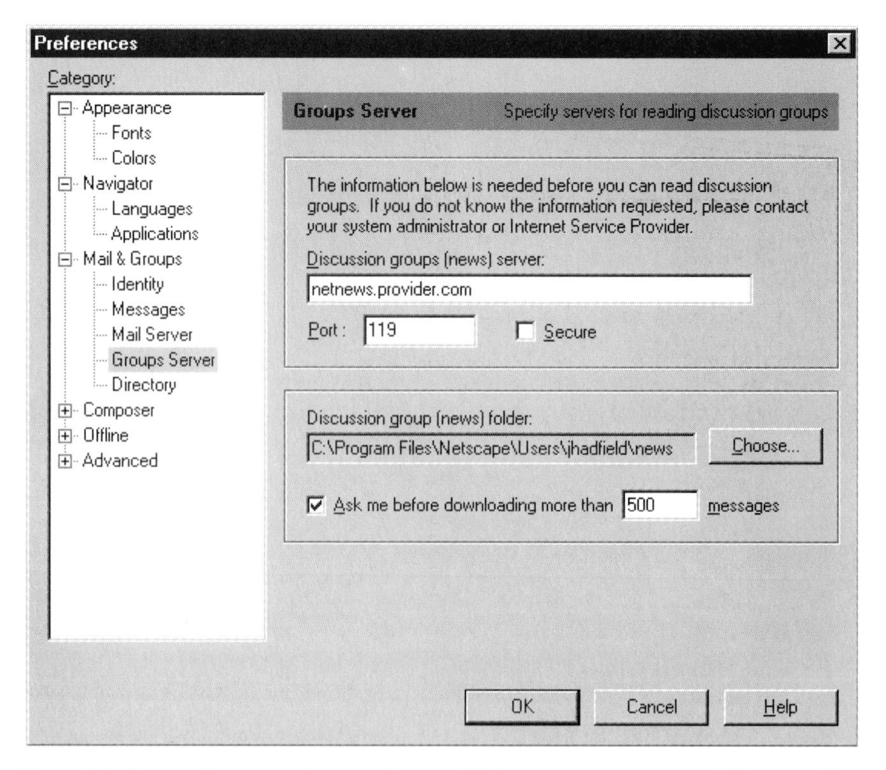

Figure 2.6 To use Netscape Communicator 4.0's Internet newsgroup abilities, you'll need to provide the name of your news server

Is This for Me?

If you're going to be the only one using Netscape Communicator 4.0 and you only need one e-mail address, just skip this section. If you want to have custom settings for different Netscape Communicator users or want to access more than one e-mail address from Netscape Communicator, read this section to get started.

Here's the basic idea. You can have only one copy of Netscape Communicator 4.0 installed on your computer. But, by using user profiles, each person using it can have different preferences, including e-mail addresses, servers, bookmarks, and more.

You've already got one user profile created for you. To create another one, exit Netscape Communicator 4.0. Then get yourself to the Windows 95 Start menu and choose Programs | Netscape Communicator | Utilities | User Profile Manager. You'll handle all your user profile stuff with this program (Figure 2.7).

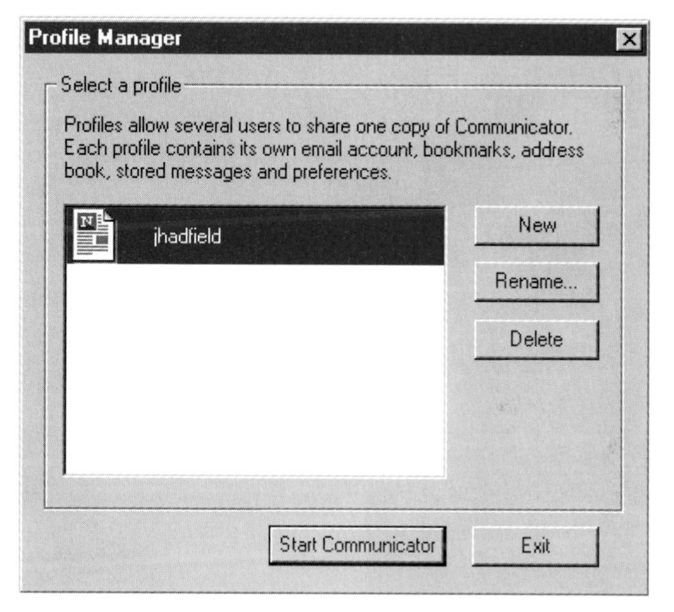

Figure 2.7 Netscape Communicator 4.0's User Profile Manager makes it easier for more than one person to use Netscape Communicator 4.0

From this snazzy app, you can add, change, or delete user profiles.

CAUTION

Each user profile takes up additional hard drive space for its own e-mail, newsgroup, and Web cache data. If you're low on disk space, you might want to avoid making very many user profiles.

When you click New, the Profile Manager Wizard will walk you through the same steps for setting up an "identity" and all the server info you entered when you first ran Netscape Communicator 4.0. Click Rename to give a user profile a new name, and, of course, click Delete to remove a profile (you'll have to confirm it before it'll really be gone).

Use Your Browser Even When Disconnected from the Net

Netscape Communicator 4.0 has a neat ability to allow you to work when you're not connected to the Net. Useful? Can be. We'll walk through it, then you can decide if you think it's easier to use it or leave it alone.

Is This for Me?

If you're connected to a network all the time, you're automatically in "online" mode—so you don't need to worry about this section. If you're using a modem to connect, or you have a laptop that's not always connected to the network, read it.

Choose Edit | Preferences, then the Offline category. Follow along onscreen or in Figure 2.8. You have three choices: Online Work Mode, Offline Work Mode, or Ask Me.

Online Work Mode assumes you're always online and doesn't prompt you to connect. Offline Work Mode will allow you to connect and disconnect from the network by choosing File | Go Online (or Go Offline). Each time you do, you'll see a special dialog box that enables you to download or send e-mail and discussion group messages:

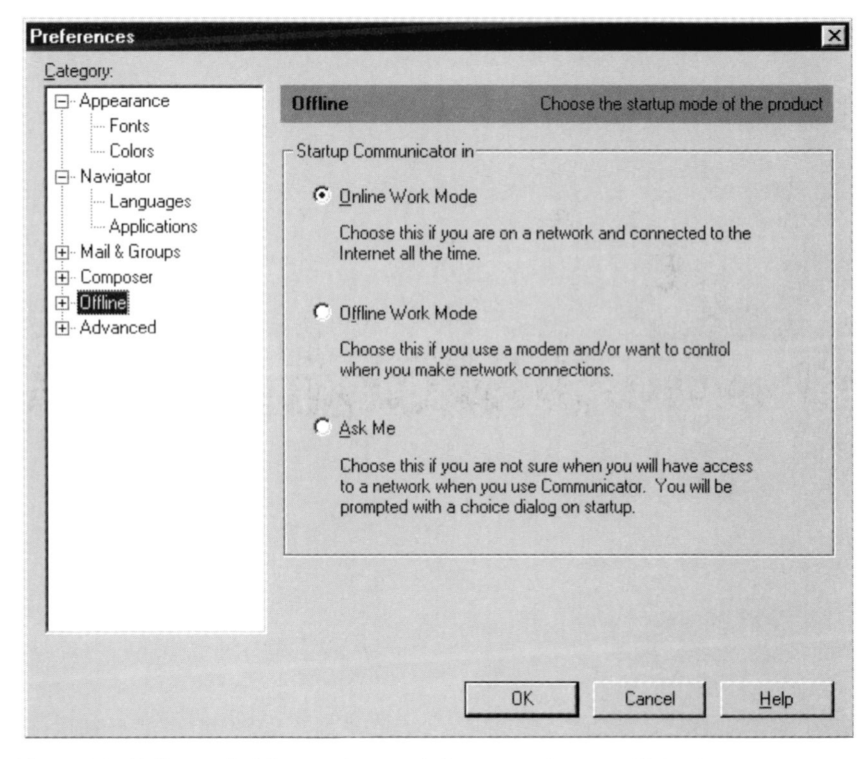

Figure 2.8 Online and offline modes can help you work more efficiently

If you're not always connected to the Net, you might want to look into Netscape's Bookmark update feature, described in Chapter 3. You may also want to explore Netcaster (Chapter 6).

Ask Me does just that—it asks you every time you start Netscape Communicator 4.0 if you want to connect to the network.

If you use e-mail frequently, it can sometimes be more convenient to download and send messages in batches. Same applies for newsgroup messages. But if you're just using Netscape Communicator 4.0 to browse the Web, these settings aren't worth fiddling with.

Now you know how to set up all the elements of Netscape Communicator 4.0—and how to connect to the proper servers and so on—to make it work just how you like it. You probably won't need to refer back to this chapter again unless you change what you like—or change Internet service providers. In the next chapters, you'll learn more about actually using Netscape Communicator 4.0 to glean information from the Internet. We'll start in the next chapter by learning how to surf the Web.

CHAPTER

3

Browsing the Web

INCLUDES

- Browsing basics
- Using URLs
- Saving pages
- Printing pages
- Turbocharging pages (turning off graphics)
- Setting/using bookmarks
- Going back in history

Understanding Web Basics ➤ pp. 52-57

Access real-time traffic information in the San Francisco Bay Area:
- PIXPage Traffic (KPIX)
- ETAK Traffic (Powered by TravInfo)
- Maxwell Traffic (Powered by TravInfo)

- The Web is millions of interrelated documents ("pages") connected via hyperlinks.
- Web pages are named using special addresses called "URLs."

Browse the Web ➤ p. 56

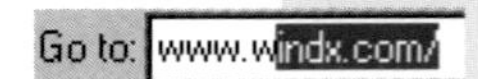

1. Type a URL (Web address) in the Location box.
2. Press ENTER (or click a hyperlink to jump to a new Web page).

Saving Pages ➤ pp. 59-60

1. Choose File | Save As.
2. Find the folder you want to save the Web page in.
3. Click Save.

Sending Web Pages via E-mail ➤ pp. 60-61

1. Choose File | Send Page.
2. Type your recipient's e-mail address.
3. Type a subject for the message.
4. Type a little note along with the automatically added URL.
5. Click the Send button.

Printing Pages ➤ p. 60

1. Choose File | Print or click the Print button on the toolbar.
2. Adjust any settings you like (number of pages to print, etc.).
3. Click Print.

Save Time by Turbocharging— Loading Pages Without Graphics ➤ pp. 61-63

☐ Automatically load images

1. Choose Edit | Preferences, then choose the Advanced category.
2. Uncheck Automatically load images.
3. Choose OK.

Add Bookmarks ➤ *pp. 63-66*

1. Right-click on a blank area of the current page.
2. Choose Add Bookmark from the pop-up menu.

Add Bookmark

Go Back in History to a Previously Viewed Page ➤ *pp. 66-67*

1. Choose Communicator | History.
2. Double-click on the page you want to jump to it.

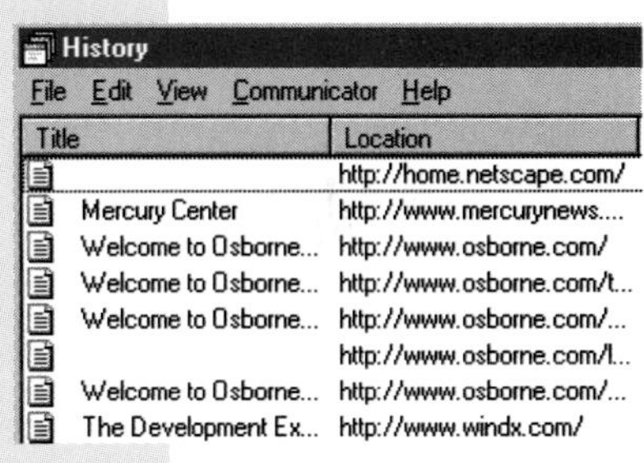

In this chapter, you'll learn the basics of Web surfing. You'll learn how things work on the Web, and you'll learn how to find what you're looking for on the Web. Most of all, you'll get a firm grasp of the basics in a matter of minutes (several minutes, but minutes nonetheless).

Is This for Me?

If you're a seasoned Web veteran, you might want to skim this chapter. But if you want a quick overview of essential info you'll need to use the Web as an informational tool, don't miss it.

Just Enough Background to Get Started

Surfing the Web isn't really that difficult. You'll need to know a few basic concepts, and then you're on your way. The first, and somewhat obvious, is that the Web is, well, a *web* of interrelated documents ("pages"). These documents are connected via hyperlinks.

DEFINITION

Hyperlinks: *Connections from phrases, words, or images on one page that jump to other pages, literally anywhere in the world. On the Web, they're usually a different color and often underlined.*

It's important to understand the idea of hyperlinking—when you click on a link, it can take you anywhere...and from there, you can link to anywhere else.

Jumping around and following links isn't anything like reading a book from start to finish. Instead, you're able to follow your whims or interests wherever they take you. For example, you might be reading a news story about Netscape Communicator. The first time the story mentions the word "Netscape," the word is highlighted as a hyperlink that connects you to Netscape's home page. At the bottom of the news story, you might see the headlines of related news stories, each its own hyperlink to the full story. (Check out Cnet's **http://www.news.com** for some examples.)

The second key concept: you need to know how you find Web pages—what their addresses consist of. These addresses are called *URLs*—and you'll learn about those next.

What's a URL, and Why Do I Care?

*You may sometimes run across URLs that don't start with "www." For example, Netscape's home page URL is **http://home.netscape.com.***

Web pages get a name—just like any other document—and an address. The combination of the two tells your Web browser where to find the page you want. This information is called a Uniform Resource Locator (URL).

You've probably already seen a URL. Here are a few examples:

- **http://www.osborne.com**
- **http://ezone.org/ez**
- **ftp://ftp.netscape.com**

Different Types of URLs

You'll run across a handful of different types of URLs for different types of Internet resources accessible through Netscape Navigator. Table 3.1 lists each type of resource, the syntax, and an example.

Is This for Me?

If you really don't care *why* you type the Internet addresses you type, go ahead and skip ahead past the table. But it won't hurt you to read a little bit about it, and we promise to be brief.

Internet Resources	How URLs Might Appear	Example
HTTP (Web)	http://www.sitename.com or www.sitename.com	**www.corel.com or http://www.osborne.com**
FTP (file archives, lists of files and folders)	ftp://ftp.sitename.com/ directory/file	**ftp://ftp.corel.com/wordperfect**
Gopher (menu-driven resources)	gopher://sitename/filename	**gopher://gopher.well.sf.ca.us**
USENET Newsgroup	news://groupname	**news://rec.music.bluenote**
TELNET (log in to a remote computer)	telnet://sitename	**telnet://spacelink.msfc.nasa.gov**

Table 3.1 Samples of Some Different Types of URLs

Before you look at the table, you might want to take a look at the elements of a URL. A URL consists of up to six elements:

```
Protocol://prefix.domain.suffix/directory/filename
```

We know—it might look confusing, but it's like junior high school algebra, except easier and without the homework. Each of the elements listed above is just a placeholder for a different type of name. For example, you might see

```
http://www.netscape.com/communicator/features.htm
```

Let's take a quick look at each of the elements.

Protocol "Protocol" is how your computer talks to others: "http://" and "shttp://" are for regular and secure Web pages, ftp:// is for file transfer, etc. Netscape assumes that the protocol for an Internet address you've typed is http:// unless it starts with an ftp prefix.

Prefix The prefix, like the protocol, is a clue to the format of any information at the Internet address. Strictly speaking, it's often not necessary to use the "www" domain prefix. But most Web addresses do—simply to make it easier to omit the http:// when they publicize the Web address.

Domain If the protocol and prefix tell your computer how it'll need to talk to the Web server, a domain is the address of the building in which the Web pages are contained (speaking metaphorically). It's the unique word by which a Web site (or FTP site, etc.) is identified. In fact, if you type only the domain name (without the prefix or suffix), Netscape Navigator will try to find it by itself and will then try adding the most commonly used prefix and suffix, www and .com.

Suffix The last part of the domain name is the suffix. You've probably seen .com the most—it's the suffix for commercial or business sites. Other common, "top-level" domain suffixes include

- .org for organizations (usually nonprofit)
- .gov for U.S. government branches
- .mil for U.S. military sites
- .edu for U.S. educational institutions

Geographic suffixes include

- .ca for Canada
- .jp for Japan
- .uk for the United Kingdom
- .fi for Finland

Directory The directory corresponds to a folder on your hard drive. In some cases, the Internet file or Web page you're looking for will be stored in a particular folder on the Web server. You won't often need to type a directory.

Filename In other cases, you'll want to specify the exact Internet document you are looking for. In those cases, you'll type the filename—it could be an HTML file or any other file format, including sounds, videos, word processing documents, etc.

STEP BY STEP Browsing the Web

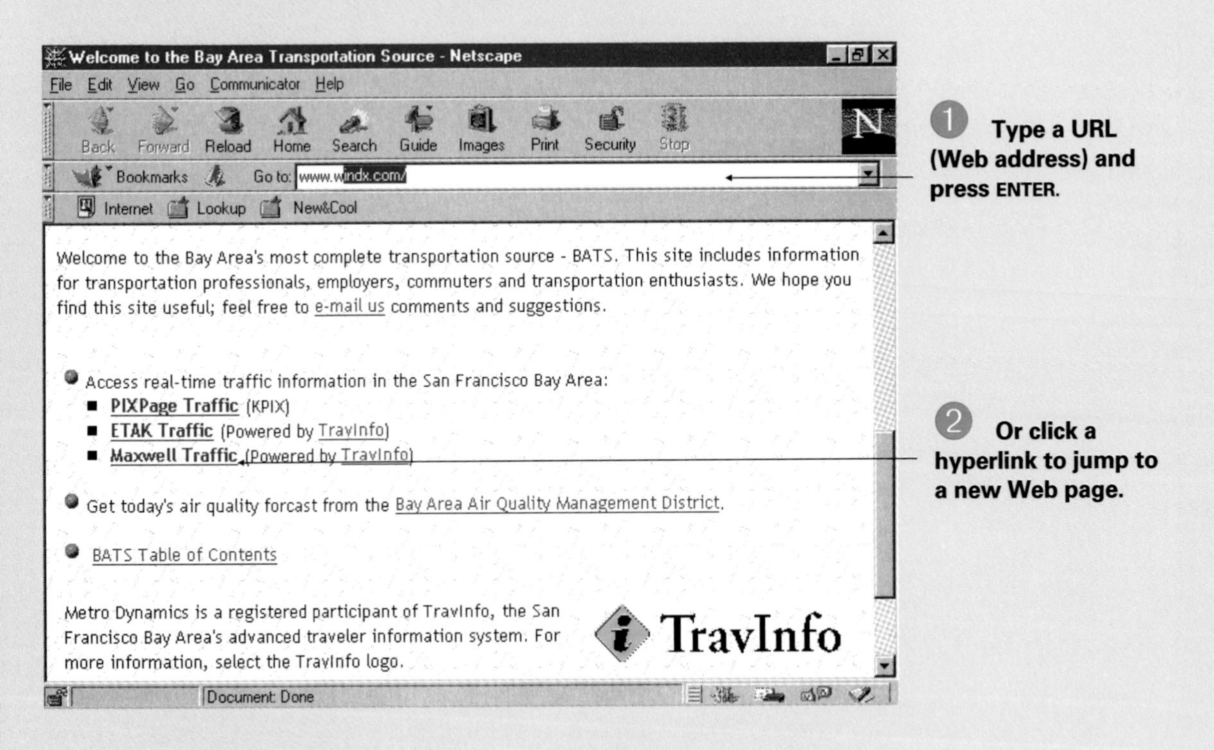

You're lucky to be using Netscape Navigator 4.0, because it includes some nifty features that make it easier to type URLs. First, you don't have to type the http:// on a Web address—as with most Web browsers, you can just type ***www.sitename.com.*** But, unlike some older browsers, you can skip the www. and .com; for example, for Netscape's Web site, you can just type **Netscape**. Netscape Navigator will infer the http://www and the .com, so, for most of your surfing, you'll save almost half your URL typing.

EXPERT ADVICE

Take advantage of Netscape Navigator's "memory." Notice how, on the browser page shown in the previous Step by Step, part of the URL is highlighted. When you type a partial Web address, Netscape Navigator will use a list of Web pages you've visited previously to automatically complete the URL for you. In this example, as we typed http://www.w, , Netscape Navigator correctly guessed that we wanted ***http://www.windx.com****—so we just pressed* ENTER.

You know, one of the nicest things about the Web is that you often *don't* have to type in long, fussy URLs. Instead, since everything's linked, you can just click and go (or make bookmarks to return to later, as discussed a little further on in this chapter).

Other Important Stuff to Know

The most important thing you can learn about using Netscape Navigator is how to use its toolbar. Here's a list of the buttons you'll see on the main toolbar—and how you'll use them.

Icon	Function
Back	Click once to move to the previously viewed page. Click and hold to view a list of previously viewed pages; then click the one you want.
Forward	If you've moved back in your viewed pages, choose this to move forward in the list.
Reload	Checks the Web server for a new copy of the current page and displays it.
Home	Returns you to your home page.

Icon	Function
Search	Takes you to Netscape's search page to help you find stuff on the Web.
Guide	Click for a menu of useful Netscape stuff such as finding people, Web Yellow Pages, new stuff, and more.
Images	(Only appears if you have images turned off) Loads or refreshes the images on the page.
Print	Prints the current page.
Security	Click here to display security information about the current page. Appears as a closed padlock if the page is securely encrypted.
Stop	Stops loading the current page.

Once you've mastered the toolbar, you'll want to learn another handy trick: right-clicking. As with other Windows 95 programs, you'll find that right-clicking on objects or links on a Web page can give you quick access to useful things. See Figure 3.1 for an example.

EXPERT ADVICE

If you find yourself using a particular feature often, check it on the pull-down menus. The menus will often list a keyboard shortcut for the feature so you can type instead of click. For example, Save As is simply CTRL-S; Back is CTRL-<.

Saving, Sending, and Printing Pages

The Web is nifty and all, but by its very nature it's transient—you're only viewing pages on the screen. But sometimes you might want to save them, share

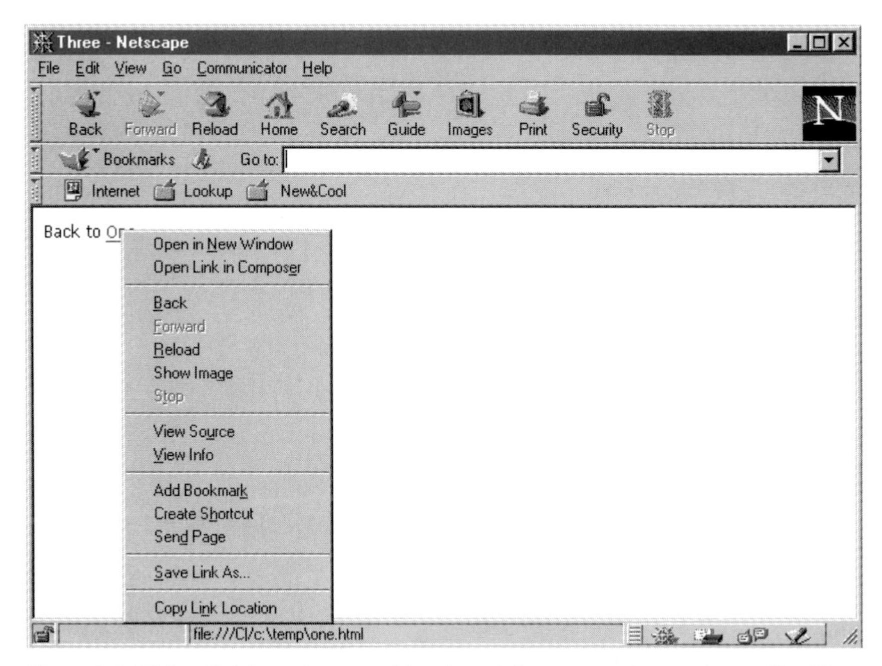

Figure 3.1 Right-click just about anything for quick access to key options—like this hyperlink

them with others, or even print them for portability. Netscape Navigator makes each of these options easy.

Saving Web Pages on Your Computer

You can save Web pages for later perusal or reference. To do so, choose File | Save As. Navigate to the folder you want to save the file in and change the filename if you choose to, then click Save.

CAUTION

Saving pages this way saves only the text portion of a Web page—not the graphics. It will save the page layout along with the text in HTML format. If you want to save only the text—in plain text format (without the page formatting), choose Plain Text from the Save As Type menu before you click Save. By using this method, you would have to save each graphic individually, if you wanted it.

Here's a trick that'll save you a lot of time: if you want to save an entire Web page—graphics and all—first choose File | Edit Page. Netscape Communicator 4.0 will open the page in its Netscape Composer module. Then choose File | Save, choose a folder and filename, and click Save.

CAUTION

Information—text, images, etc.—may well be copyrighted, even when it's distributed on the Web. Most copyright laws allow you to save things for personal use, but if you're planning to distribute them, get permission first.

Sending Pages via E-mail

You've just run across some great info on Java (the programming language) and want to send it to your mom. Rather than printing it and sending it in the mail, you can just e-mail it to her (your mom has e-mail, right?). See the Step by Step instructions on the next page to find out how.

Print Web Pages for Later Perusal

As backwards as it might sound, you'll probably print a lot of your favorite Web pages. Face it: your computer screen is not the most comfortable setting for long or involved reading sessions. If a Web document is too long, or if you want it handy in a more familiar form, try printing it. You may be surprised at how good some Web pages look that way (even if they are a little long, usually, for a single piece of paper).

In fact, it was rumored that Michael Kinsley, editor of Microsoft's Web-only magazine *Slate*, printed early issues and mailed them to friends in New York City so they'd read the material in a familiar medium.

It's easy to print a page: just choose File | Print or click the Print button on the toolbar. You'll then see the Print dialog box—you can customize which pages you print, choose how many copies, and more. When you've adjusted the options, choose Print.

STEP BY STEP Sending a Page via E-Mail

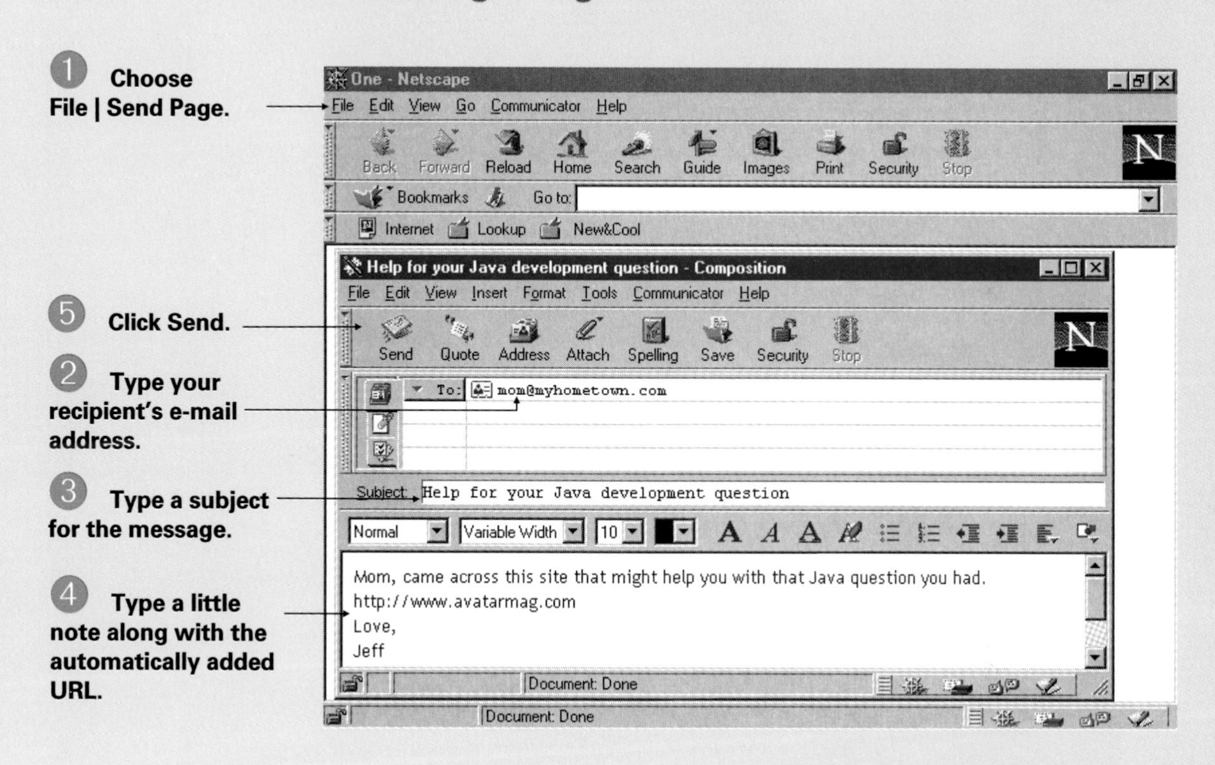

Turbocharge Pages: Turn Off Automatic Image Loading

Using a modem? You can accelerate your page's load time with this nifty trick. (Many hardcore Web surfers swear by this: they can find more information without waiting for graphic niceties.) Netscape Navigator can load Web pages containing text—with placeholder graphics (Figure 3.2) instead of the full-fledged graphics. Admittedly, they're ugly, but they load fast.

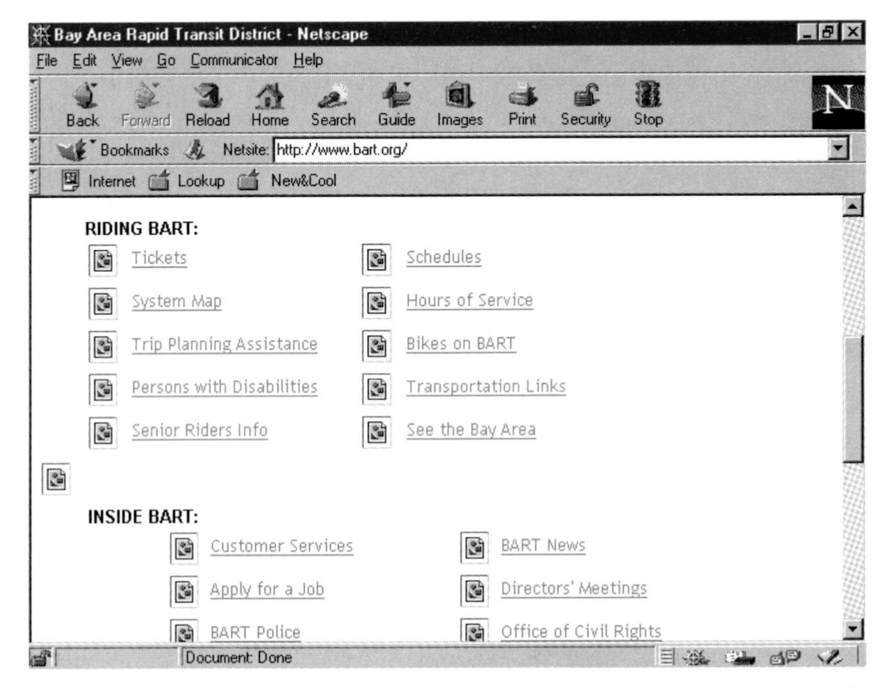

Figure 3.2 If you've decided to view your pages without graphics, they'll be replaced by placeholder icons.

EXPERT ADVICE

Many sites provide "low-bandwidth" or text-only pages. If you're bookmarking pages, set your bookmarks to the text-only or low-bandwidth pages. They'll load faster and generally provide the same information and abilities without the glitz.

If you feel you're missing out and want to load the graphics, just click the Images button on the toolbar.

Here's how to turn images "off."

1. Choose Edit | Preferences, then choose the Advanced category.

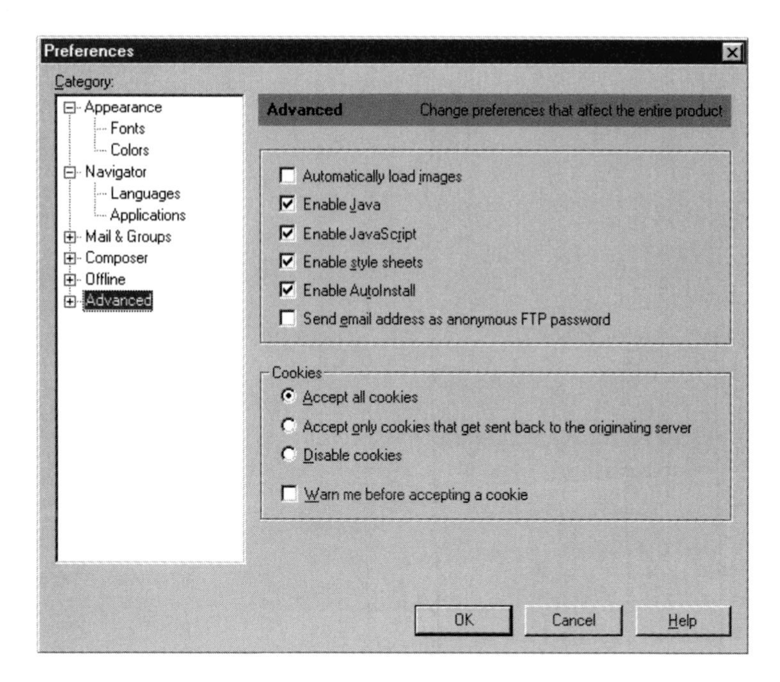

2. Click to uncheck Automatically load images.

Bookmark Sites You'll Want to Visit Again

Want to save your place so you can come back to it again? Didja find a Web site or Web page you *know* you're going to find indispensable? Bookmarks are your answer. In fact, they're so easy and useful that *everyone*, and I mean, *everyone*, uses them. They're one of the best features of Web browsers.

They're called bookmarks because they're like that ticket stub you slip in your paperback book: they hold your place so you can come back to it later. But they're even more like a personal address book of favorite places to visit, because your bookmarks are yours and yours alone—you can edit them, search through them, or just browse among the entries.

Netscape Communicator 4.0 makes keeping bookmarks easier than ever—and you can place them into folders and organize them, too.

The quickest way to set a bookmark is to right-click on a blank part of the Web page, then choose Add Bookmark from the pop-up menu:

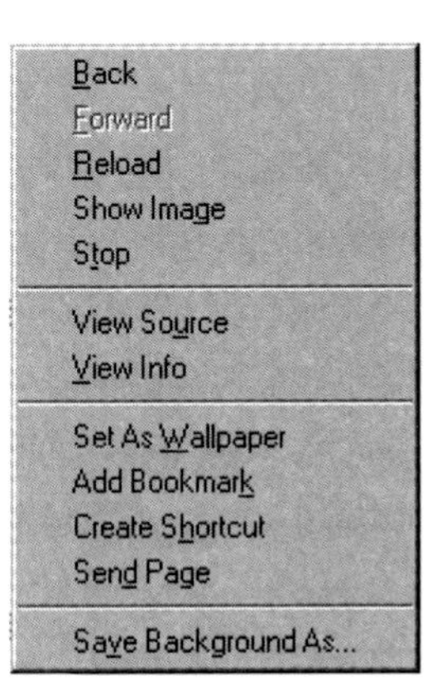

EXPERT ADVICE

Just "adding" bookmarks is like saving treasures in a junk drawer. You know where they're stored, but you have to rustle around to find what you want. You're better off putting things where they belong—filing your bookmarks in particular folders.

But this method is like dumping your socks wherever is most convenient rather than putting them away in your sock drawer. You may want to use the slower-but-wiser way and *file* your bookmarks as you save them.

SHORTCUT

To create the bookmark folders in the first place, press CTRL-B or choose Communicator | Bookmarks | Edit Bookmarks. Choose File | New Folder to create folders, and drag and drop bookmarks into folders to organize them.

STEP BY STEP Filing Bookmarks

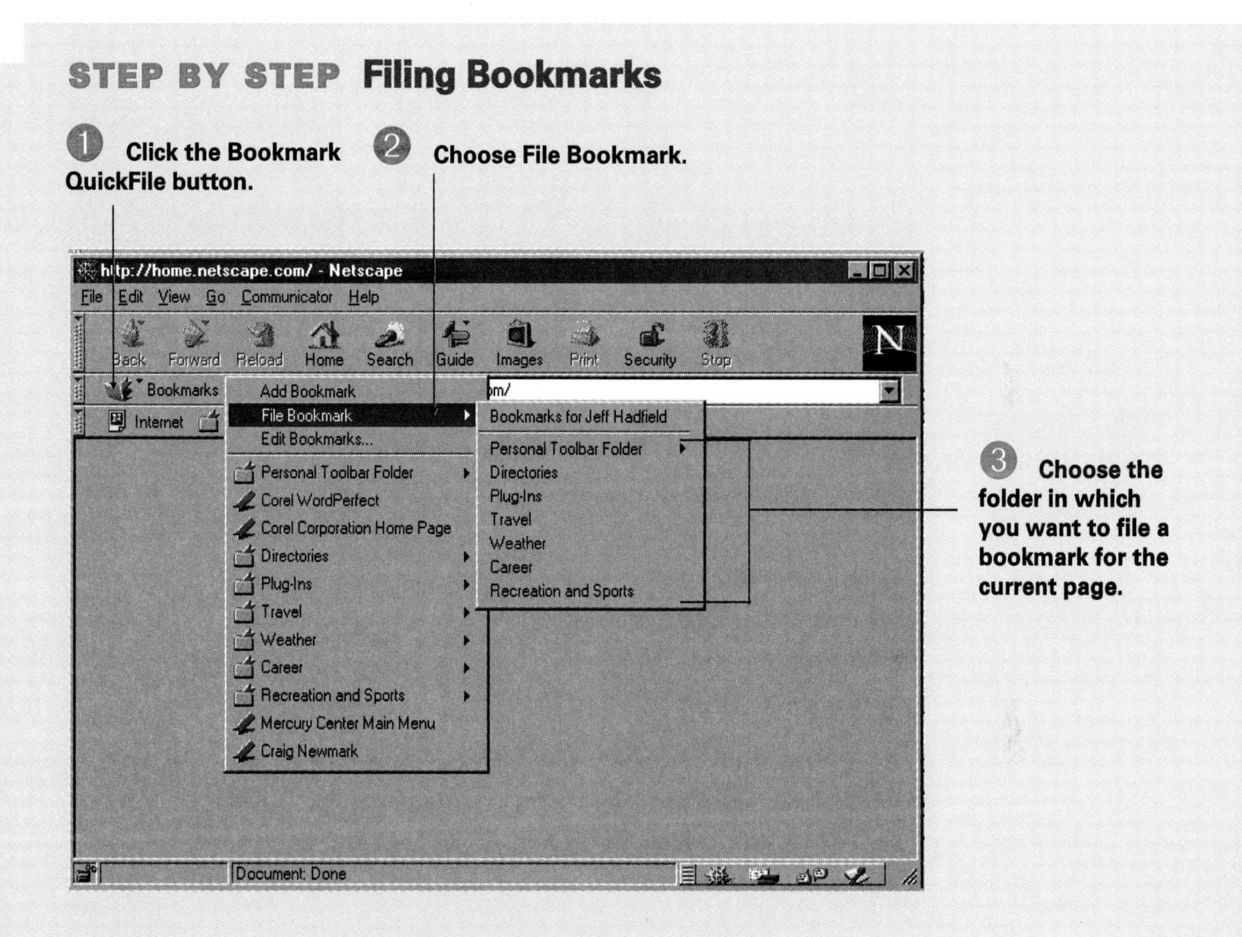

Checking for Page Changes

Netscape Communicator 4.0 has a cool new feature that can alert you when one of your bookmarked pages has changed. This saves you the effort of checking on each of your favorite sites for something new. To try it out, first bring up the bookmark list by pressing CTRL-B or choosing Communicator | Bookmarks | Edit Bookmarks.

You'll see your default bookmark folder. If you only want to check a few of your bookmarks, highlight those (you can select multiple bookmarks by holding down the CTRL key while you're clicking on each one). Then, choose View | Update Bookmarks:

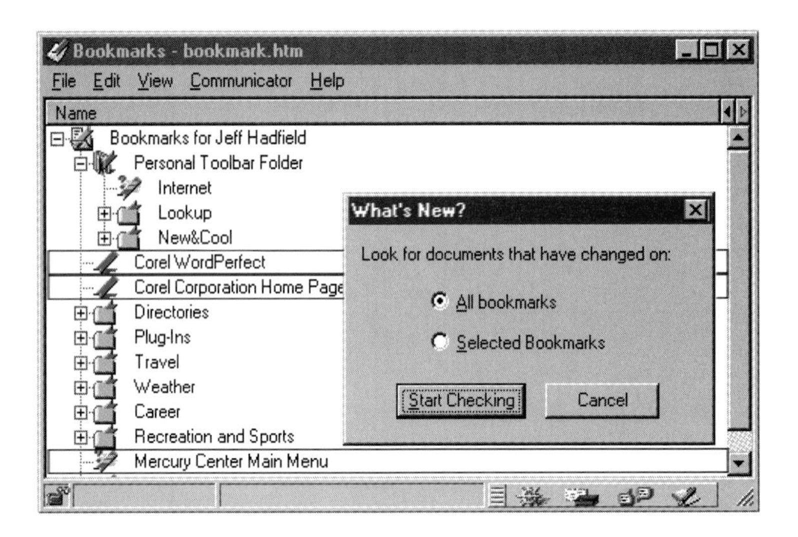

If you want to check all of your bookmarks for updates, choose All bookmarks; to check only the ones you've highlighted, choose Selected Bookmarks. Choose Start Checking, and Netscape will show you its progress as it searches for changes. When it's done, the bookmark list will show updated sites with a highlighted icon (if it couldn't tell if the site was changed, it'll list it with a question mark).

Going Back in History

Like Bill and Ted , you can go back in history to visit pages you've recently visited. You've already learned about using the Back button and the Back button menu, but these only let you return to pages you've visited since you ran Netscape Communicator 4.0 *this* time—during the current "session."

Use the History feature to view a list of recently visited documents. Choose Communicator | History or press CTRL-H. You'll then see the history window shown in Figure 3.3.

History

File Edit View Communicator Help

Title	Location	First Visited	Last Visited	Expiration	Visit Count
	http://home.netscape.com/	7/6/1997 ...	7 days ago	8/5/1997 8:53 PM	20
Mercury Center	http://www.mercurynews....	7/22/1997...	7 days ago	8/5/1997 8:32 PM	7
Welcome to Osborne...	http://www.osborne.com/	7 days ago	7 days ago	8/5/1997 9:00 PM	2
Welcome to Osborne...	http://www.osborne.com/t...	7 days ago	7 days ago	8/5/1997 9:01 PM	2
Welcome to Osborne...	http://www.osborne.com/...	7 days ago	7 days ago	8/5/1997 9:01 PM	2
	http://www.osborne.com/l...	7 days ago	7 days ago	8/5/1997 9:02 PM	3
Welcome to Osborne...	http://www.osborne.com/...	7 days ago	7 days ago	8/5/1997 9:02 PM	3
The Development Ex...	http://www.windx.com/	7 days ago	7 days ago	8/5/1997 8:52 PM	1
The Development Ex...	http://www.windx.com/scr...	7 days ago	7 days ago	8/5/1997 8:53 PM	2
	http://www.windx.com/scr...	7 days ago	7 days ago	8/5/1997 8:53 PM	2
	http://www.windx.com/fre...	7 days ago	7 days ago	8/5/1997 8:53 PM	2
	http://www.windx.com/scr...	7 days ago	7 days ago	8/5/1997 8:53 PM	2
	http://www.windx.com/scr...	7 days ago	7 days ago	8/5/1997 8:54 PM	1
	http://www.windx.com/scr...	7 days ago	7 days ago	8/5/1997 8:55 PM	1
	http://www.windx.com/scr...	7 days ago	7 days ago	8/5/1997 8:56 PM	1
	http://www.windx.com/scr...	7 days ago	7 days ago	8/5/1997 8:57 PM	1

Netscape

Figure 3.3 Draw upon this History list of pages you've viewed—and sort it by clicking the buttons at the top of each column

See Chapter 2 to learn about setting the History feature's length—from one day to several weeks.

Double-click on any page listed to jump to that page. You can sort the list by clicking the column headings—just click the column you want to sort by (clicking toggles between listing them first to last and last to first).

EXPERT ADVICE

Recently visited Web sites are also listed under the Go menu. You can return to sites you've visited since you started Netscape Communicator 4.0 from this menu.

Searching Within a Web Page

You will learn more in Chapter 4 about finding things on the Web. But what if you want to find a phrase within the current Web page? Netscape Navigator makes that easy, too.

1. Choose Edit | Find in Page. You'll see the Find dialog box.

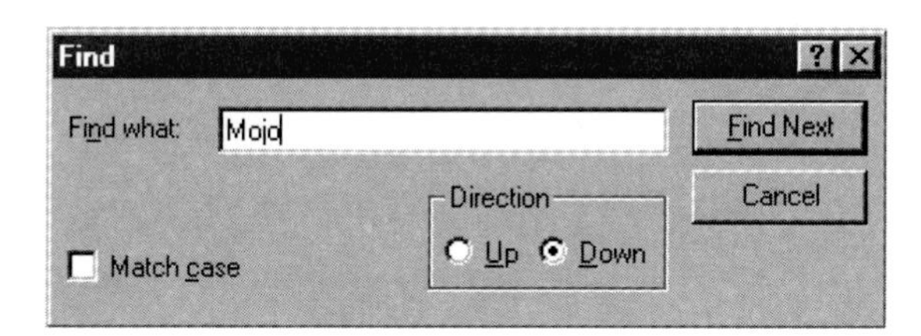

Want to learn how to search the Web itself? Saunter over to Chapter 4.

2. Type the word or phrase you want to search for.
3. If you want to search for the exact case (upper or lower), check here.
4. Choose the direction in which you want to search.

Now that you've read this chapter, you should be familiar with the basics of Netscape Navigator. If you haven't done it already, get out there and surf the Web for a few minutes to try out your new skills. Once you've done that, move on to the next chapter to learn how to find things on the Web.

CHAPTER

4

Finding Things on the Web

INCLUDES:

- Finding stuff on the Web using Netscape's search
- Search directly from Netscape's Go to box
- Finding stuff using other search sites
- Downloading files
- Using downloaded files
- Finding someone on the Internet

Find Stuff on the Web ➤ pp. 72-75

1. To search the Internet for information, click the Search button on the toolbar.
   You'll then be connected to Netscape's search page.
2. Type the phrase or word you want to search the Web for and choose Search.
   Netscape will send your search phrase to the search engine and return the results on a Web page.
3. Scroll through the search results and click the hyperlink to the Web page that contains the information you're looking for.

Search Directly from the Go to Box ➤ pp. 75-76

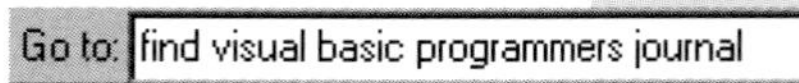

1. In the Go to box (where you'd usually type **http://*www.sitename.com***), type **find**, then press SPACEBAR.
2. Type the keywords for which you want to search.
3. Press ENTER.

Find and Download Files ➤ pp. 78-82

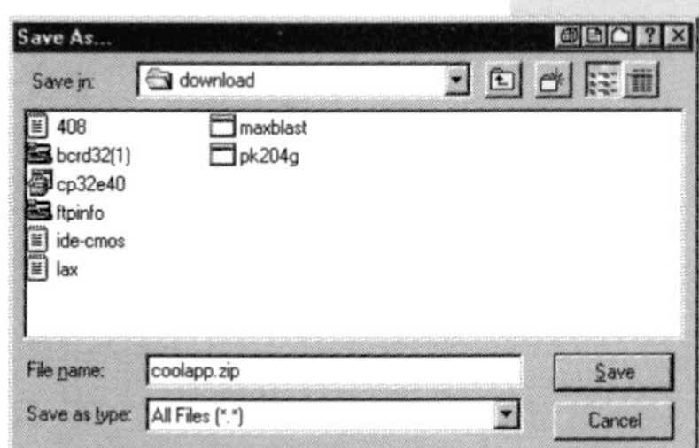

1. Find the file you want using a reputable Web site such as **http://www.download.com** or **http://www.shareware.com.**
2. Click the highlighted link or the download button for the file you want. You'll see Netscape's Save As dialog box.
3. Use the Save in drop-down menu and choose the directory you want to save the file in.
4. If you want to change the name of the downloaded file—although usually, you won't need to—type that name in the File name box.
5. Choose Save to begin downloading the file to your hard drive.

Use Downloaded Files ➤ *pp. 82-83*

- Look for information about using the file on the Web page you downloaded it from.
- Watch for files in two formats:
  - Self-extracting executables, which automatically uncompress themselves
  - Compressed archives, usually in Zip format (or StuffIt on the Mac)
- Follow any instructions included in the archived/compressed file; they'll usually be in a "readme.txt" file.

Find Someone on the Internet ➤ *pp. 83-86*

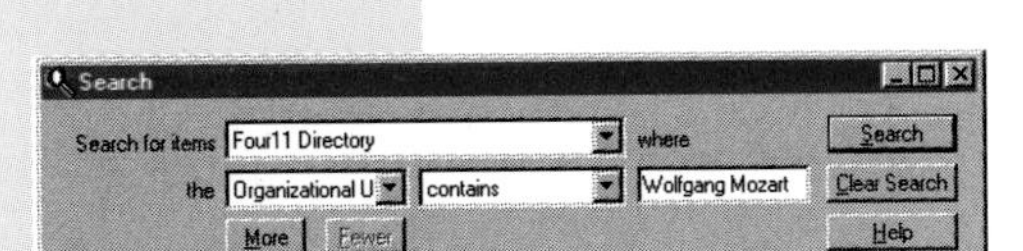

1. Choose Edit | Search Directory.
2. If you want to try a different directory, choose it from the directory pull-down menu.
3. Choose the information you want to use as your starting point from the Name pull-down menu.
4. Choose the word that best describes how you want to look for this information (usually *contains* or *sounds like* work best).
5. Type the actual information you want to use as your starting point in the unlabeled text box.
6. Click Search.

The Web contains grundles (that's a unit of measure larger than both a mess and a passel) of information. Hundreds, even thousands of new Web sites appear every day. There's so much information available that it's not humanly possible to know where every possible bit of information might be.

Some of the most popular services on the Web help you with this very problem. They're designed to make it easy to find the information you want. (Of course, the degree to which they succeed varies.)

That's what you'll learn in this chapter: how to find information, Web sites, and even other people using Netscape Navigator and the Web. You'll start out by using Netscape's search page, and then you can take a look at some of the larger search engines and their Web pages. Finally, you'll learn how to find people, phone numbers, and e-mail addresses online.

Is This for Me?

We honestly can't think of anyone who *won't* need to look for something on the Web at some point. Even if you don't need to find anything right now, you might want to skim this chapter. When you do need to find something, you'll know to return to this chapter to see how.

Finding Stuff on the Web

Most of the time, you'll know that you just want to find some stuff on the Web about some subject. Not a very specific request, sure—but it's a start.

For example, perhaps your child has been diagnosed with reflux esophagitis. Like most of us, you've probably only the vaguest idea of what that means—why not use the Web to find out for sure? Or, say you're planning a biking vacation to the Pacific Northwest. You might want to find others who have done so, magazine articles about the subject, and travelers' tips for cycling. Or maybe you're just looking for some reviews of some old Audrey Hepburn movies you've been meaning to rent.

Netscape's Built-in Search Abilities

Netscape's got your answer built right in. To begin any search for information, start here. To search the Internet for information, click the Search button on the toolbar:

You'll then be connected to Netscape's search page, which will look similar to the one shown in Figure 4.1.

Netscape has done a good job of collecting the major search resources into one place for you. (Of course, you can visit any of these sites on your own.) The default search engine varies each time you access the page; for us, it's often Excite.

DEFINITION

Search engine: *A tool you can use to find information on the Net. Each search engine's company claims it's the best, but each can be useful.*

If you'd prefer to use a different search engine, click a tab (shown in Figure 4.1) to choose the one you want. You can choose from Yahoo, Excite, Infoseek, and Lycos. We'll talk later about deciding which search engine will work best for you.

Type the phrase or word for which you want to search the Web, and choose Search:

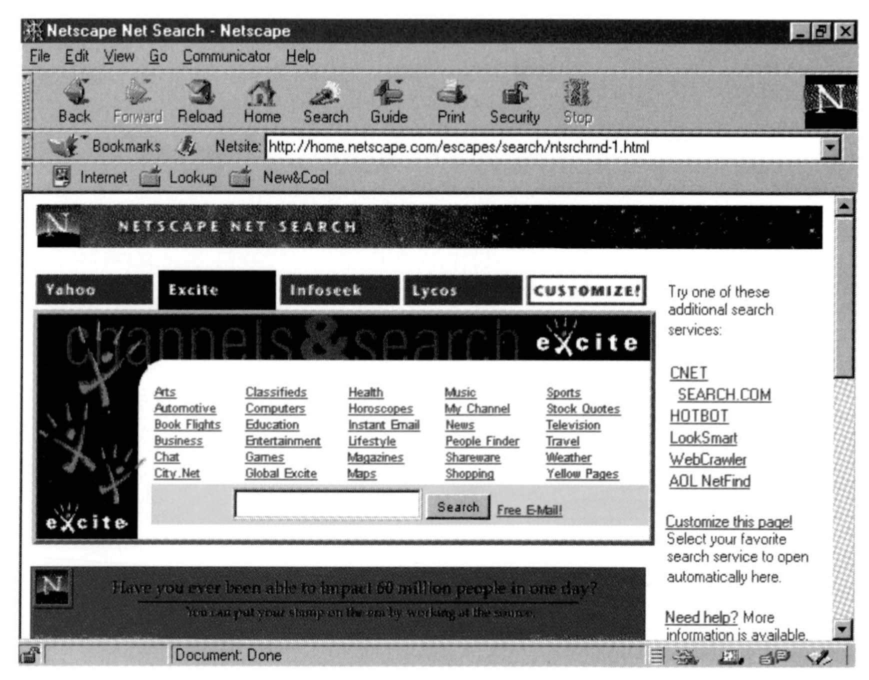

Figure 4.1 Netscape's search page. From here you can search or link to any of the major search resources on the Web

Netscape will send your search phrase to the search engine and return the results on a Web page like the one shown in Figure 4.2.

You'll see a list of Web sites. Often only ten or so will be listed per page, but most search engines will contain a link to allow you to see the next sites in the list (see Figure 4.3). Each site in the list contains a hyperlink you can click on to jump directly to that Web page. You'll also see a description or brief summary of each to help you decide which sites you want to view.

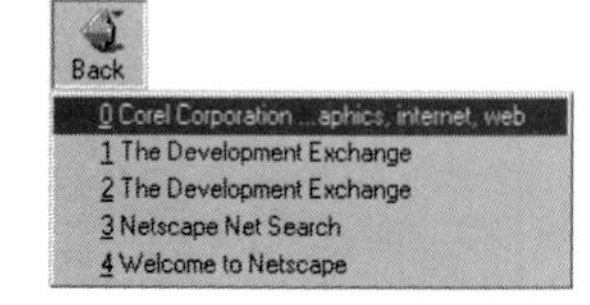

Click a link to see if a listed Web page contains the information you're looking for. When you're ready to return, click the Back button until you return to your search results page. Or, you can click and hold the Back button until a list of previously viewed sites appears, then choose the most recently viewed Web search page (the one highest on the list).

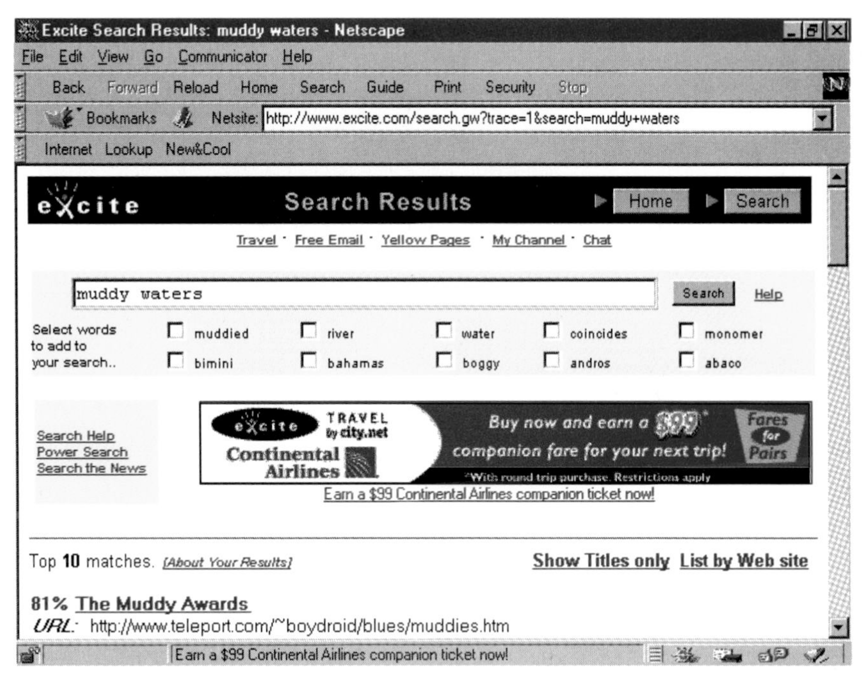

Figure 4.2 Results of a search. Click the highlighted links to jump to a Web page that looks like a likely candidate

SHORTCUT

Once you've found a site that seems well-tailored to the information you want, be sure to look for a links page there. Most sites include a list of other useful related sites. These lists are sometimes more useful than search engines because they're compiled by people who know the subject matter very well.

Quick Searches

If you're just not in the mood to hassle with a lot of options, you can search directly from the Location | Go to box. Instead of a URL, just type **find** and then

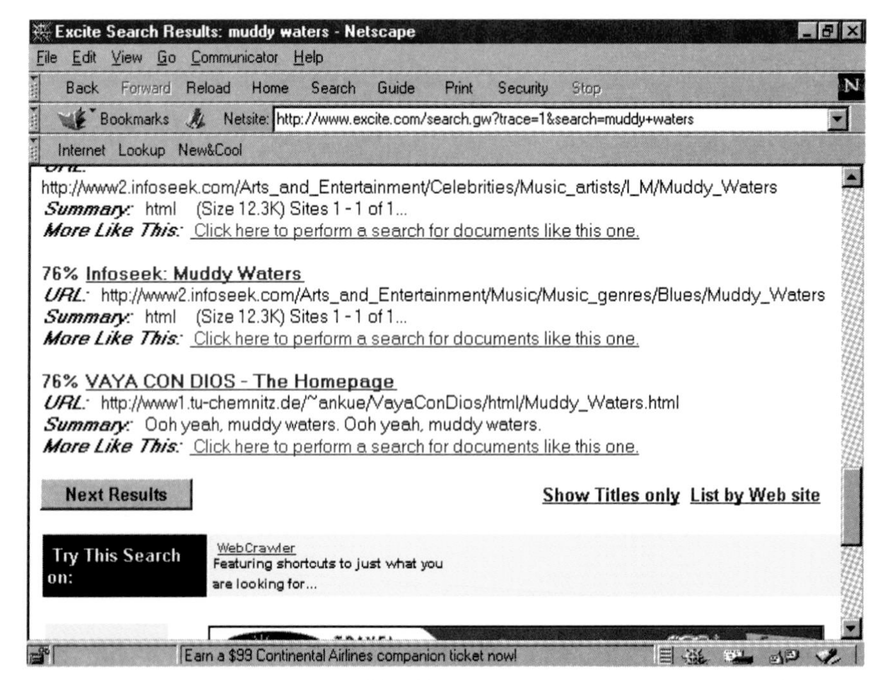

Figure 4.3 Most search engines—like this one—give you easy ways to scroll through a long list of search results

your search keywords. Your search will use Netscape's default search engine, which, for us, is usually Excite.

To search instantly:

1. In the Location | Go to box (where you'd usually type **http://*www.sitename.com***), type **find**, then press SPACEBAR.
2. Type the keywords you want to search for.
3. Press ENTER.

You'll see a search results page, as described earlier.

Which Search Engine?

These Web search engines are really the only way you'll ever find anything you're looking for on the Web. Sure, sometimes it'll be easy: if you're looking for advice about diapers, you might start with http://www.pampers.com. But in most cases, you can't guess the URL or location of the information you want.

Enter search engines. Each approaches the Web differently, each uses its own technology (or human staff) to catalog and index the Web, and each is useful in its own way. Netscape's search page is pretty cool because it provides a one-stop shop for the major search engines.

EXPERT ADVICE

Most search engines allow you to use something called Boolean operators. While this sounds almost creepy, it really just means that you can use words like AND, OR, and NOT in your search terms. If you type a list of terms, it's understood that you want to search for ***fish*** *(and)* ***trees*** *(and)* ***Cleveland****. But you can be more specific by saying that you want to search for* ***fish and trees not Cleveland*** *to return all results that list fish and trees—but specifically don't list Cleveland. By the same token, OR helps you use synonyms—similar words—as in,* ***trees or forest or foliage****.*

Each search engine has strengths relative to the others. For example, Yahoo uses a large staff to catalog and describe Web sites. You can browse or search for its listed Web sites, each listed in the category from which it came. In this way, Yahoo acts somewhat as a table of contents for the Web. Excite, on the other hand, has only recently begun to place its listings in categories. Instead, it (and many others, like AltaVista and HotBot) uses technology to automatically index and describe Web sites. Different from all of these is DejaNews, which archives postings from Internet newsgroups and then allows you to search those archives.

If you're truly serious about searching for specific information on the Web, it's worth searching with a few of the different engines. However, remember: with these powerful search engines, it's very easy to get *more* information than you want or need, so choose the search engine that you're most comfortable using and explore from there.

EXPERT ADVICE

Here's a search strategy that works well for us: Search both Yahoo and Excite. If you still don't find what you want, try HotBot. AltaVista is unmatched for brute power—but it can easily return much more information than you want, so choose your search terms carefully.

Finding and Downloading Files and Shareware

You're not just looking for information on the Web: it's also a great resource for files—including

- New drivers and updates for your computer hardware
- Shareware
- Graphics and images
- Fonts and typefaces
- Drivers, updates, and patches for your computer's software

EXPERT ADVICE

A nifty product from Cybermedia, called Oil Change, automatically searches the Web for updated hardware and software drivers for your system. It can also help you install and maintain these updates. As the name implies, it's a relatively easy way to keep your system tuned up and operating at its peak (you could search for all the drivers on your own, but why bother?). Trend alert: Microsoft is building similar abilities into Windows 98 as part of the operating system (there'll be a Windows Update icon on the Start menu).

Finding the Files You Want

Like finding other information on the Web, sometimes it's amazingly easy and other times it's frustratingly difficult to find the files you want. Numerous Web sites host vast collections of downloadable software. You might want to start with these:

Each of the search engines contains listings of popular and useful download sites.

- **http://www.download.com**
- **http://www.shareware.com**
- **http://www.tucows.com**

Downloading Stuff

The neat thing about finding files on the Web is that you can download them straight from Netscape Navigator. When you click on a link in a Web page that links you to a downloadable file, you'll see the window shown in Figure 4.4.

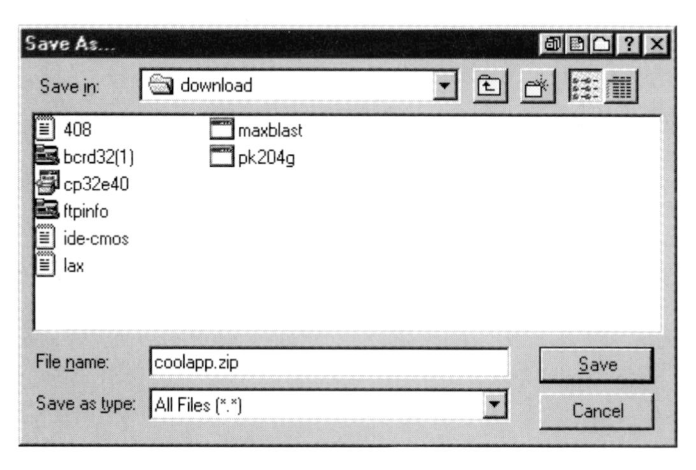

Figure 4.4 Navigate to your special download folder (if you like), then choose Save to download a file

Netscape Navigator doesn't care what format the file's in—so you'll need to make sure you know what to do with it once you've downloaded it (we'll talk about that in a moment). But first, here's what you do once you see the window in Figure 4.4. (If you've ever saved in another Windows 95 application, you'll be familiar with this procedure.)

1. Use the Save in drop-down menu and the folder icons to choose the directory you want to save the file in (See "Keeping Track of Downloaded Files," below).
2. Usually you won't need to, but if you want to change the name of the downloaded file, type that name in the File name box.

CAUTION

Be sure to name the file with the same three-digit file extension (the part after the period). Windows gets very confused if you don't. For example, if the file was named coolapp.zip, you'd want to rename it okayapp.zip.

3. Choose Save to begin downloading the file to your hard drive.

Now, it may take a while for your file to download, depending on its size. Many well-designed Web sites tell you about how long they expect a file download to take depending on the speed of your Internet connection. In any case, once you've started the download, Netscape will keep you abreast of the progress:

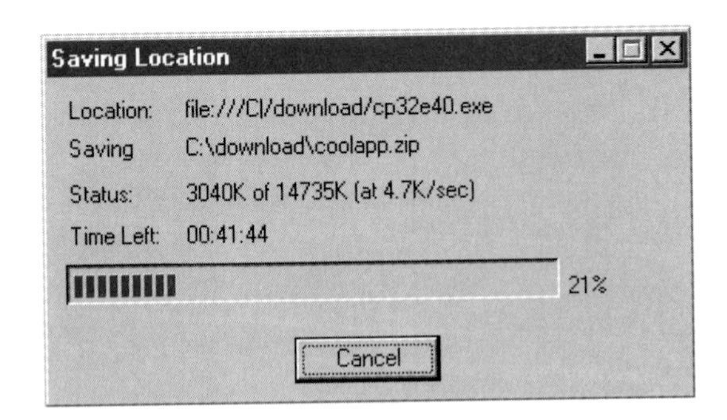

Keeping Track of Downloaded Files

As your mother taught you, it's best to have a place for everything and keep everything in its place. This applies equally well to your downloaded files. We

EXPERT ADVICE

Here's where modern technology makes us happy: you can do other stuff while you're downloading the file. You can go back and browse the Web, read your e-mail, or whatever—just click back on the Web browser window. The file download will automatically close that window when it's done. But don't be fooled: especially on a modem, you'll notice that your other Web-related tasks are slower because Netscape's still downloading your file in the background.

strongly recommend you download all your files to a specific folder. You might move them later, but if you always download to the same folder, you'll always be able to find them.

Our recommendation is admittedly not particularly creative, but you'll thank us later: create a file called "download" on your hard drive. Here's how:

1. From your Windows 95 desktop, double-click My Computer.
2. In your My Computer window, double-click your hard drive icon.
3. Choose File | New | Folder.
4. A new folder icon will appear in the window named New Folder. Type **download** and press ENTER.

Be Smart and Safe

It sounds like a public service announcement to tell you to be smart and safe, but you'll want to do two key things when you're downloading programs from the Internet:

- Watch for viruses
- Download reputable programs

Watch for Viruses

Invest in virus protection software and run it regularly. This software will keep your system scanned for the telltale signs of viruses and help you keep your computer free from infestation. You may even want to scan files after you've downloaded them and before you run them, just to be safe.

Providers of antivirus software provide good instructions for running and maintaining their software. Two favorites: Symantec's Norton Antivirus and McAfee's VirusScan.

DEFINITION

Virus: *An insidious program, sometimes deliberately malicious, that can at least annoy you—and, at worst, destroy your software and data.*

EXPERT ADVICE

Whether or not you have virus protection software, back up your important data regularly, too. Sorry to preach—but if you've backed up your data, the damage any virus, software, or hardware glitch can cause is diminished.

Trust Your Download Site

Something you download is only as trustworthy as the site you've downloaded it from.

Our most important advice: download programs only from companies and sites that you trust. These might be major software publishers—like Netscape, Microsoft, Corel, Symantec, or others—or trusted hardware manufacturers—like Dell, Compaq, Gateway, Sony, IBM, or others. Or you may feel comfortable downloading software from publishers of major magazines or Web entities like CNET. In any case, if you're unsure about the source of software, treat it warily.

What to Do with It Once It's Downloaded

Often, the page from which you download a file includes information about using the downloaded file. You can print that page and save it for later reference.

However, most downloaded files are compressed, often in PKZip ("zip") format for Windows or DOS (.sit—StuffIt—for Macintosh). Many files will be saved in what are somewhat painfully termed "self-extracting executable archives."

In other words, just run the file you downloaded and it will extract itself. If it's not in this format—if it has a "zip" extension—you can extract the files from the archive using a utility such as PKZip for Windows or WinZip (see Appendix A for sources for these utilities).

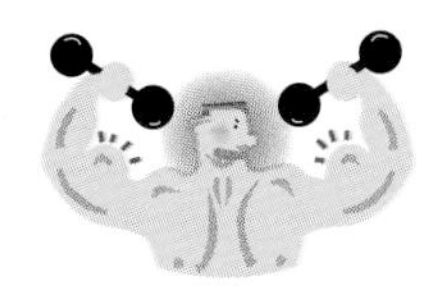

DEFINITION

Compressed files: *Files saved in a special format that makes the file size as small as possible for distribution or archiving.*

How Do I Find Somebody?

It's cool but scary at the same time: almost all U.S. phone book information is accessible on the Web. And there's more information available than just phone numbers and addresses—in many cases, you can find people's e-mail addresses, too. Privacy issues aside, it's quicker than dialing 411 on the phone.

You can go to great lengths to find people in any of the several Web-based directories (Four11's our favorite, at **http://www.four11.com**). But Netscape Navigator makes it so you don't have to remember all the site's names and URLs. It has integrated major Web directories right into its Search Directory feature.

Give it a try: you can search for just about anything (phone numbers, e-mail addresses, etc.). To find people on the Web:

1. Choose Edit | Search Directory. You'll see the Search Directory window:

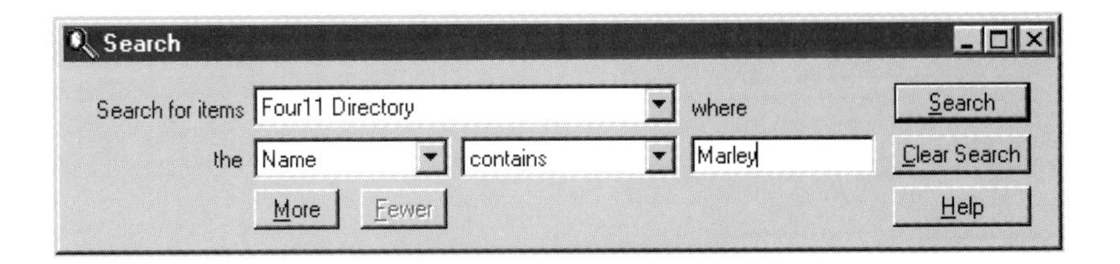

You can search any of several directories, but Four11's a good one to start with.

2. If you want to try a different directory, choose one from the directory pull-down menu:

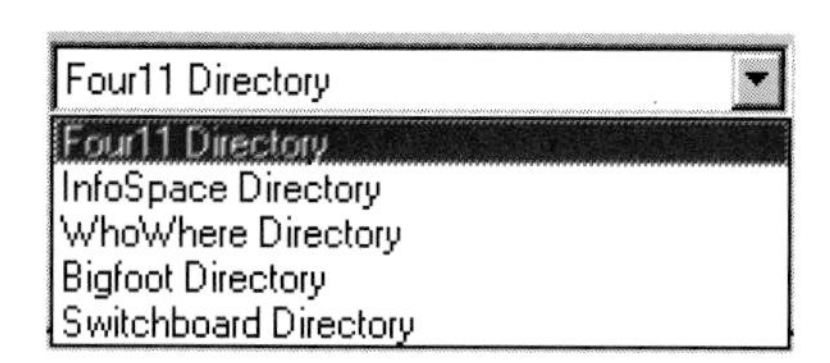

3. Choose the information you want to use as your starting point from the Name pull-down menu:

EXPERT ADVICE

If you want to specify more than a simple search, click the More button. You'll get an additional line on which to specify more search criteria (choose Fewer to make the extra line go away). You can add as many lines as you like. For example, you might have a friend named Wolfgang Armadillo Mozart who lives in Albuquerque. You could search for him by specifying that you want information about someone whose: name contains Wolfgang Mozart, name does not contain Amadeus, city sounds like albekerkee.

You can choose to start your search using:

- Someone's name
- Someone's e-mail address
- Someone's phone number
- The name of a business or other organization

- The URL of a business or other organization
- The name of a city
- The name of a particular street

4. In the next pull-down menu, choose the word that best describes how you want to look for this information. Most of the time, you'll stick with "contains." However, you might want to choose "sounds like" from the menu if you're searching for a city name of "San Jose," in which case you would type **san hosay**.

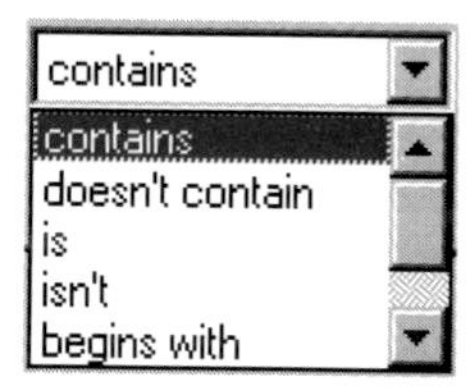

Here's a quick look at how each of these options works:

- *Contains* allows you to type all or part of a word that's in the name, e-mail address, or other information you're looking for.
- *Doesn't contain* lets you type all or part of a word that *isn't.*
- *Is* and *is not* let you specify exactly what is and isn't part of the information you're looking for.
- *Begins with* and *ends with* are pretty easy to figure out: if you know exactly how the information you're specifying starts or ends, you can use these.
- *Sounds like* is a cool way to ensure that misspellings won't get in the way of your search: it looks for any thing that even resembles the word you've typed.

EXPERT ADVICE

In most cases, you can stick with contains, *the default choice, or* sounds like. *Most of the time, it's just as fast to search with less specific information—and then to narrow your search once you see the results.*

5. Whew! Final step: type the actual information you want to use as your starting point in the unlabeled text box:

6. Click Search.

   When your search is complete, you'll see a window like the one in Figure 4.5 (it'll vary depending on the search directory you chose). Depending on the search directory, you may need to click the hyperlinked name you want in order to see the complete information.

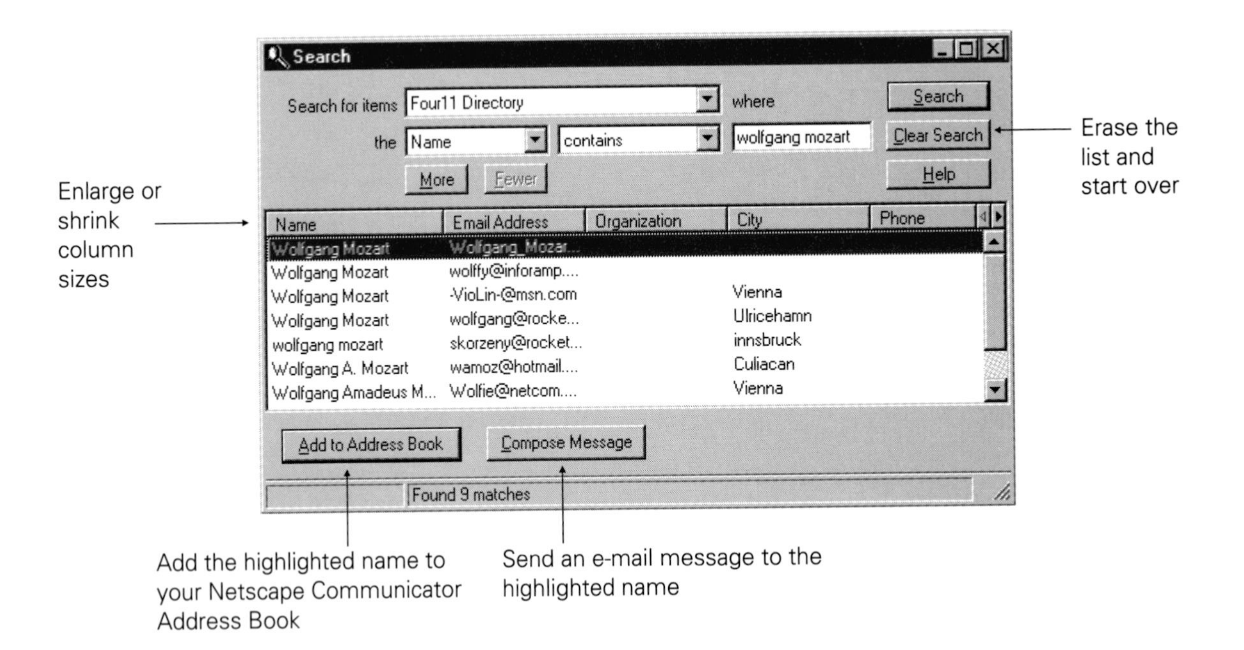

Figure 4.5 A completed search. Highlight the name you want, then click one of the buttons

In this chapter, you've learned how to actively find information, files, and people on the Web. In the next chapter, you'll learn how to have the Web come directly to you—with a lot less effort than you've had to expend to find stuff.

CAUTION

CHAPTER

5

Instant Information with Netcaster

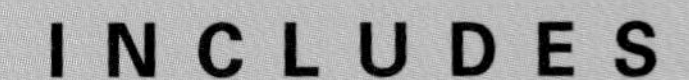

INCLUDES

- What will Netcaster do for you?
- Starting and closing Netcaster
- Choose channels
- Choose a view: Webtop vs. browser
- Setting correct preferences

FAST FORWARD

What's Netcaster? ➤ pp. 92-95

Netscape Netcaster is a new way to regularly automatically download information from the Web and display it on your computer—without having to go hunt for it. This information can be displayed in a regular Netscape Navigator window or in the background as a Netcaster Webtop.

What's a Webtop? ➤ p. 94

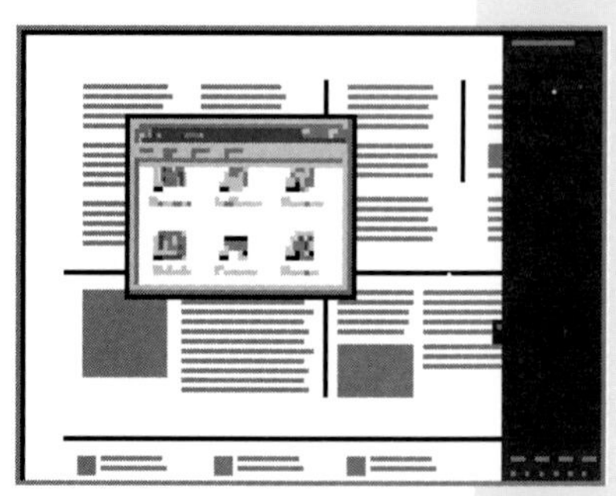

A Webtop is a full-screen Web page that appears in the background of your screen—like Windows 95 wallpaper. Unlike wallpaper, however, it's updated as often as you like with the site's latest information.

Start Netcaster ➤ p. 95

If you're already running Netscape Communicator, just choose Communicator | Netcaster. Or, from the Windows 95 Start menu, you can choose Programs | Netscape Communicator | Netscape Netcaster.

Use the Channel Finder ➤ pp. 96-99

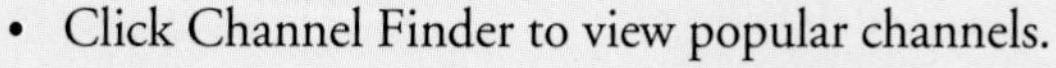

- Click Channel Finder to view popular channels.
- Click the topic tabs to view different groups of channels.
- Click More Channels to view a longer list of channels.
- Click My Channels to view your favorite automatically updated channels.
- Click any channel name for a brief preview.
- Click Options to change Netcaster settings.
- Use the bar to adjust Webtop display.

Add Channels to Your List ➤ pp. 99-102

Your list of My Channels includes channels you want automatically updated and displayed.

1. In the Channel Finder, click the channel name you want to add.
2. In the channel preview, click the Add Channel button.
3. Adjust any options you like in the Channel Properties dialog box.
4. Click OK.

Control Your Netcaster Webtop Window ➤ pp. 102-104

Use the Webtop toolbar to control your Netcaster Webtop. Among other things, you can

- Move among pages
- Move among Netcaster channels
- Move the Webtop to the front or back of your display
- Print the current page
- Close the Webtop

Fine-tune Netcaster ➤ pp. 104-106

1. In the Channel Finder, click Options.
2. Under the Channels tab, you can add, delete, and modify channels.
3. Under the Layout tab, you can move the Netcaster Channel Finder between the right and left sides of the display, automatically hide the Channel Finder, and more.
4. Under the Security tab, you can adjust advanced security settings.
5. Choose OK to close the dialog box.

Netscape's newest addition, Netcaster, turns your PC into a $2,000 television—only it's not *quite* as flexible. Like a television, Netcaster delivers channels of information you can sit back and view at your leisure. Unlike a television, however, Netcaster's information isn't being *broadcast* to you as it happens. Instead, it consists of information that's been downloaded to your hard drive—you can review it at *your* convenience, rather than the broadcaster's.

Netcaster is one of the first applications to enable what Internet types call "pushed" information: instead of you needing to request, or "pull" something, all you need to do is make your information needs clear one time—then it's delivered to your computer as often as you like (see Figure 5.1 for a sample).

This "pushed" information is delivered in channels, much like your television channels. For example, you might want business information from the CNN and Wired channels, along with some information for your real life from the HomeArts channel.

DEFINITION

Netcaster channel: *A Web site copied to your desktop. Each channel consists of information formatted just like regular Web pages using HTML and other technologies.*

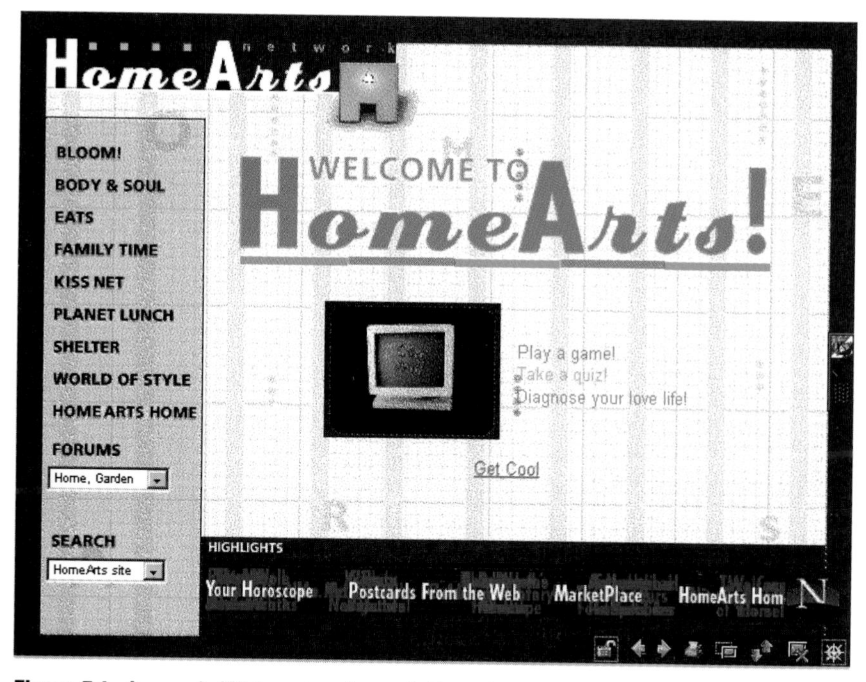

Figure 5.1 A sample Webcaster channel. Note the small button on the right side of the screen—click there to use the Channel Finder

What Can It Do for You?

Netcaster can save you time—but, ironically, it's by far the slowest of Netscape Communicator's components. Its greatest benefit is its biggest drawback.

Netcaster saves you time because it enables you to predownload information you want. In effect, you're priming the pump, telling Netscape you want to view certain Web sites, you want it to check for new information on those Web sites, and you want a certain level of detail predownloaded from those sites.

But, since Netcaster is built on some fairly new technologies, it's unfortunately annoyingly slow to set up and launch. Once it's running, its performance is reasonable. But unless you're running on a zippy machine, you may find its performance issues outweigh any time you might save by predownloading information.

Is This for Me?

Netcaster is some nifty technology. But it's only useful if you're truly interested in integrating often-consulted Web information into your desktop, and, most importantly, if you're running a machine with a fast Internet connection (faster than 28.8 kbps), a fast processor, lots of RAM (at least 16MB), and a whole mess of free hard disk space. If you have the time, give it a try. If you don't feel like taking the time to set it up, adjust it, and experiment with it, just bookmark your favorite sites in Navigator and visit them often.

Speed issues aside, Netcaster can be useful to the busy person. It'll download information you want in big clumps—so you can browse the information even when you're not connected to the Net. (One user reports that it helps avoid Internet traffic jams during peak hours, too.) And Netcaster can display its information in the background of your other Windows—what Netscape calls a *Webtop.* Webtops overlay your existing desktop—much as one of those large desk calendars sits on top of your physical desk. The Webtop displays your Netcaster information, complete with updates, so that you can glance at it when you want to. For instance, the screen shown in Figure 5.1 is a Webtop—it obscures everything else on the screen (but sending the Webtop to the background of your screen is only a few mouse clicks away).

EXPERT ADVICE

*You may have heard of Microsoft's similar plans. Microsoft has integrated Internet Explorer 4.0—their Web browser—with Windows 98 to create what they call Active Desktops. These are much like Netscape's Webtops, except that they're integrated with the operating system. Microsoft's format for the "channels" they provide is different than that of Netscape's. However, the two companies are expected to support each other's formats—mostly to make it easier for the companies providing information for the different channels. Our take: Netscape's Webtops remain a viable alternative for those who prefer Netscape's interface, those who don't want their Web browser integrated into their operating shell (or display), and those who run operating systems other than Windows 98. Both Netscape and Microsoft are building on foundations built by the pioneer in the field, PointCast (**www.pointcast.com**).*

Starting Netcaster

You can start Netcaster on its own or from within a different module of Netscape Communicator. If you're already running Netscape Communicator, just choose Communicator | Netcaster.

Or, from the Windows 95 Start menu, you can choose Programs | Netscape Communicator | Netscape Netcaster.

Running Netcaster for the First Time

The first time you launch Netcaster, you'll see a Java security warning (Figure 5.2). Check the Remember this decision box, then click on Grant to allow Netcaster's broadcast channels to send you information.

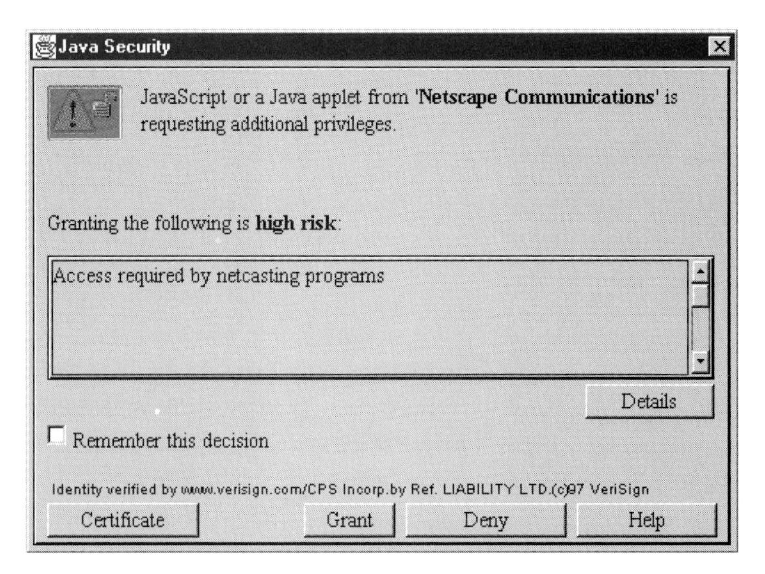

Figure 5.2 When you launch Netcaster for the first time, you'll see this security warning. Check the Remember this decision box, then click on Grant

CAUTION

While it's okay to grant this access to your computer when it comes from Netscape or Netcaster content providers, it's not always okay. Java applets run just like other programs on your computer—and malicious ones couid cause damage. Allow Java applets only from trusted sources. If you doubt the source, click on Certificate to view the site's credentials.

As Netcaster launches, you may see similar Java security warnings. Click on Grant for each of these, checking the Remember this decision box each time so you won't be bothered again.

Once you've started Netcaster, a large window will appear as the background of your screen and the Channel Finder will appear on the right side (see Figure 5.3).

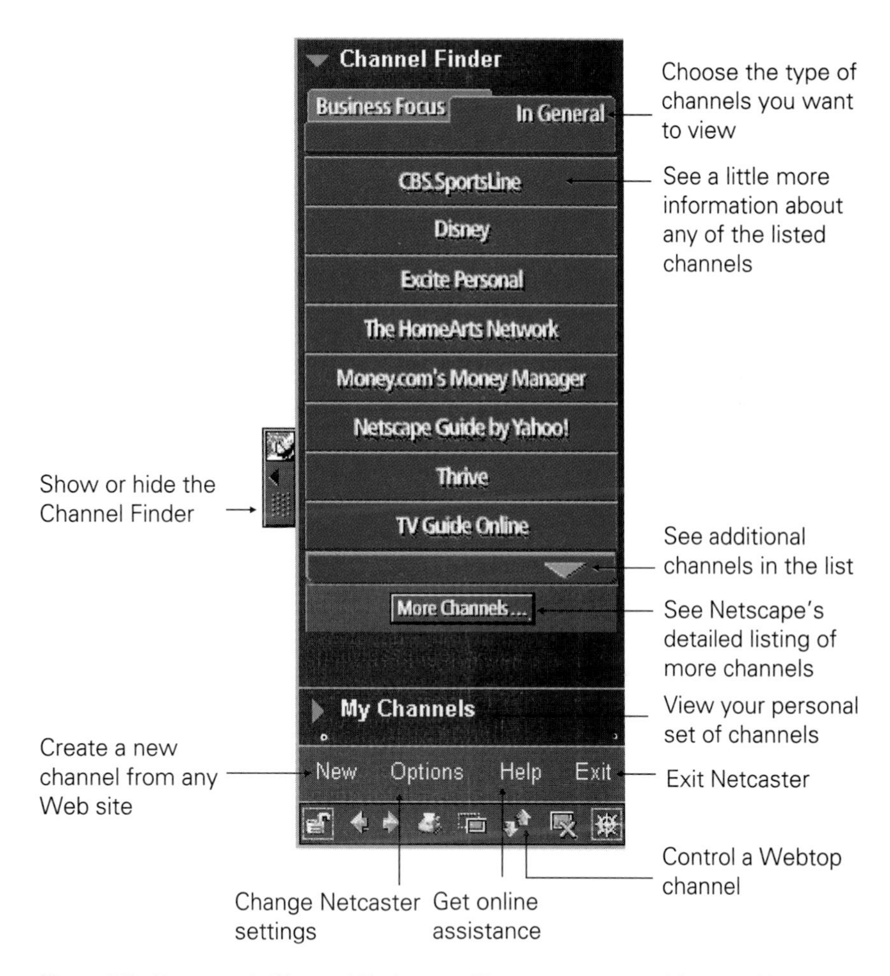

Figure 5.3 Netcaster's Channel Finder acts like a remote control for your Netcaster window

EXPERT ADVICE

Due to the graphical nature of the Netcaster channels, Netscape recommends you leave automatic image loading turned on (Edit | Preferences | Advanced). If you've got it turned off, Netscape will warn you the first time you run Netcaster.

What Are All These Buttons?

When you buy a new television set, you're befuddled at first by the new remote control (and, sometimes, you remain befuddled with those that are poorly designed). Netcaster's the same way: its Channel Finder window is confusing at first. Look at Figure 5.3 for some help.

Choosing Your Favorite Channels

When you run Netcaster for the first time, you have the opportunity to select just what information you want it to download, how much of it to download, and other custom settings. You can change these settings anytime, but you'll want to set up some initial stuff.

One of the most fun and useful things to do is to choose your favorite channels, and Netcaster will update them for you automatically. You'll find channels in two places: in the Channel Finder listing and in Netscape's listing of More Channels. Your favorite channels will be placed in the My Channels list.

You've already seen some of the Channel Finder. When you click a channel in the list, you'll see a small graphic preview of the channel plus two buttons—one allows you to preview the channel, the other allows you to add the channel to your My Channels list.

EXPERT ADVICE

Netscape keeps the Channel Finder up-to-date, so you may want to check it now and then for new channels.

Once you've added a channel to your list of My Channels, it'll appear like this in your Channel Finder:

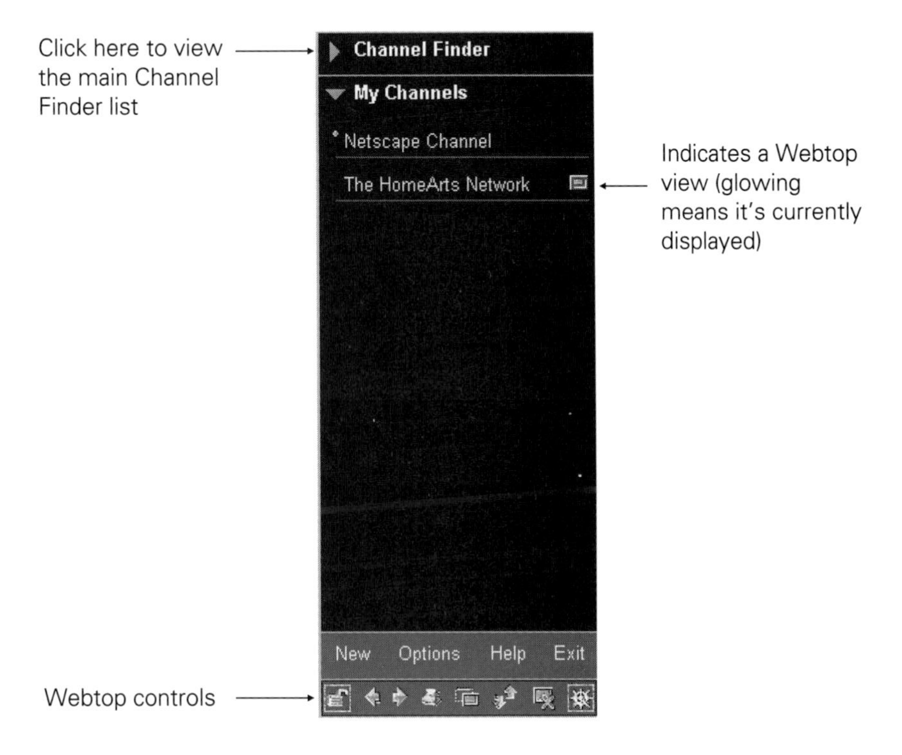

Try it now: find a channel you like in the Channel Finder. Click the channel name, then click Add Channel.

EXPERT ADVICE

The first time you add a channel, you'll probably have to register with Netscape's Member Services by filling out a short, informational survey—doing this is required if you want to download channels.

Add a Netcaster Channel to Your My Channel List of Favorites

Surprisingly, adding a channel to your list of favorites is pretty easy. Mind you, we've listed each step in detail below—but you'll find the process straightforward.

1. Change the channel name or location, if you want—usually you won't mess with these, though.
2. Check here to have Netscape automatically update the Netcaster site's information for you.
3. Choose how often you want the site updated from the pop-up list.

4. If you've chosen a daily update or the like, choose the time or day you want the page updated.

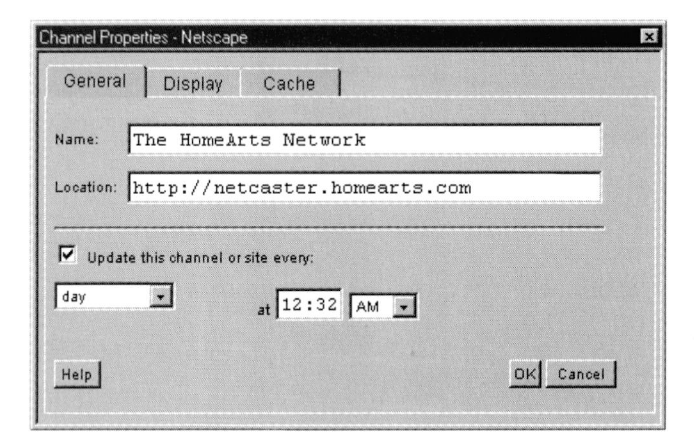

5. Click Display to adjust more settings.
6. Choose to display the channel in a regular Navigator window—like a regular Web site—or as a Webtop in the background of your desktop.

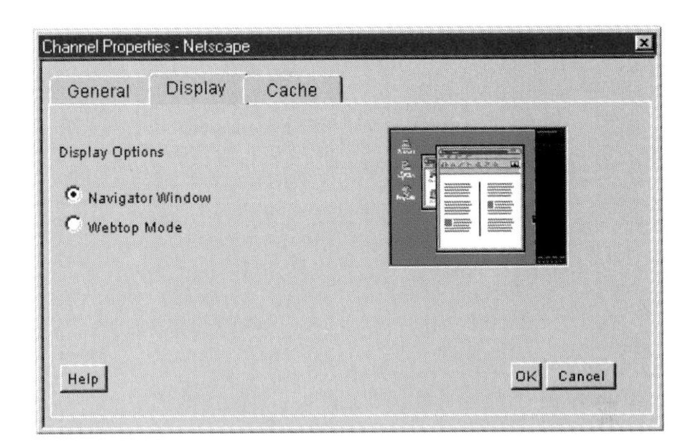

7. Click Cache to adjust more settings.
8. Choose the number of levels in the site you want to download automatically. Imagine the site as a tree—or organizational chart—of linked pages. Remember: you can always view additional levels by connecting to the Internet.
9. Choose the maximum amount of disk space—in kilobytes—you want the channel to use on your hard disk drive. For example, 6000K would correspond to about 6MB.
10. Click OK to add the channel to your list of My Channels.

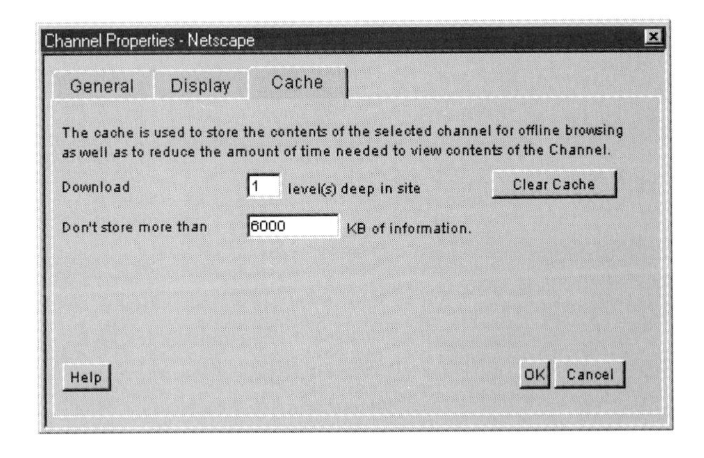

Some channels (CNN Financial for one) allow you to set some special options specific to those channels. More channels in the future will probably incorporate these, so watch out for them.

EXPERT ADVICE

You can add channels from Netcaster's options window (shown later in this chapter), as well. The options shown in the Add Channel dialog box are the same as the ones you've just learned—except that you'll have to specify a channel name and a URL.

Webtop View vs. Navigator View

Netscape touts the Webtop view as their latest and greatest addition. In many ways, it is. It creates a dynamic backdrop to your applications that's updated as frequently as you like with information you want.

However, if you're interested only in viewing Web sites without waiting for them to load, you might simply ask Netcaster to display your My Channels choices in regular Navigator windows. You'll probably just be browsing the Netcaster sites as you would any other Web site, so it'll be more convenient to navigate in the Navigator view.

Is This for Me?

Why should you choose a Webtop view? It's a nice way to have essential information always available. It appears in the background and doesn't require any additional effort to update. However, it can make your computer display visually busy and difficult to read. If you're not already adept with working with windows, switching among applications, and the like, it'll probably cause more confusion than help. Stick with the standard Navigator view if you're leery.

You can switch between display styles on any channel you've chosen. To change the display style of a Netcaster channel:

1. Choose Options from the Netcaster Channel Finder window.
2. Under the Channels tab, highlight the channel you want to change.
3. Click Properties.
4. Click the Display tab.
5. Choose the display style to which you want to change.
6. Click OK, then Close.

Remember that on a Netcaster channel's Webtop display, you can still click on hyperlinked words or images.

If you've chosen Webtop view, you'll find some handy tools for getting around. These tools, shown in Table 5.1, always appear in the bottom right-hand corner of the screen.

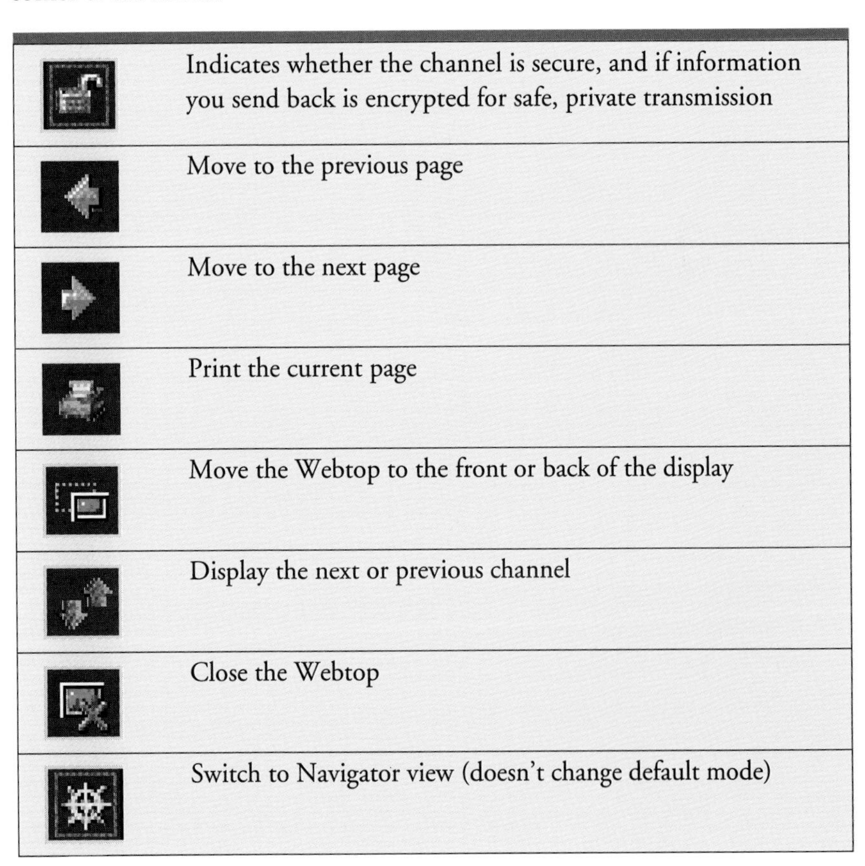

	Indicates whether the channel is secure, and if information you send back is encrypted for safe, private transmission
	Move to the previous page
	Move to the next page
	Print the current page
	Move the Webtop to the front or back of the display
	Display the next or previous channel
	Close the Webtop
	Switch to Navigator view (doesn't change default mode)

Table 5.1 Use the omnipresent Webcaster buttons to control your Webtop

From the Channel Finder, you can adjust several fine-tuning options. Click Options in the Webcaster control panel to see this window:

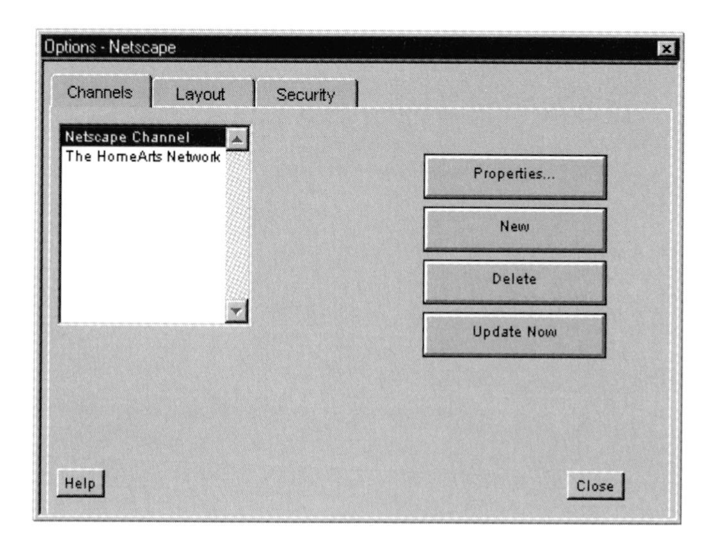

Fine-tuning Netcaster

While you've learned about tuning each individual channel, you can also fine-tune Netcaster itself to work just the way you like it. To do so, choose Options from the Channel Finder.

Once you see the Options dialog box, you can change any number of Netcaster settings.

1. Under the Channels tab, highlight any channel you want to change.
2. Click Properties to change channel settings (just as you did when you initially added the channel to your list of My Channels).
3. Click New to add a channel. You'll need a name and the URL for the Web site you want to use. The rest of the settings are just the same as adding a channel from the Channel Finder.
4. Click Delete to permanently remove the channel from your list of My Channels.
5. Click Update Now to download the latest information for the channel regardless of its update settings.
6. Click the Layout tab to adjust more settings.

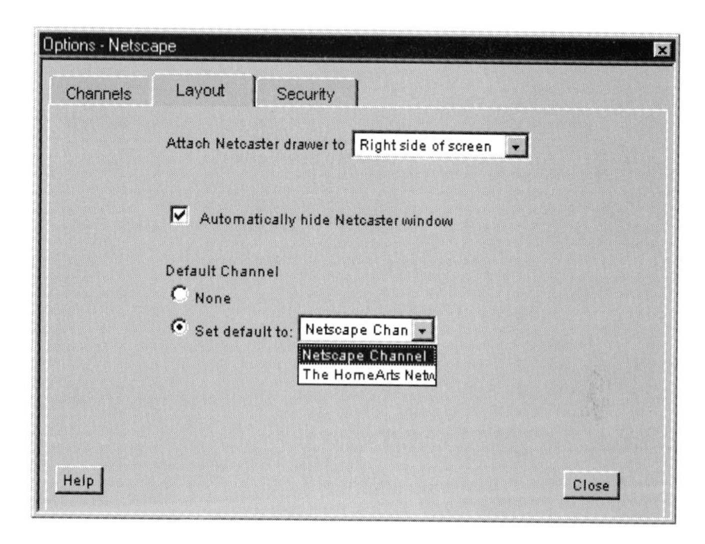

7. In the Attach Netcaster drawer box, choose either the right or left side of your screen.
8. Check here to automatically hide—or "close"—the window whenever a channel is displayed.
9. Choose None to launch Netcaster with a blank window—or choose the channel you want it to load automatically from the pop-up list.
10. Click the Security tab to change more settings.

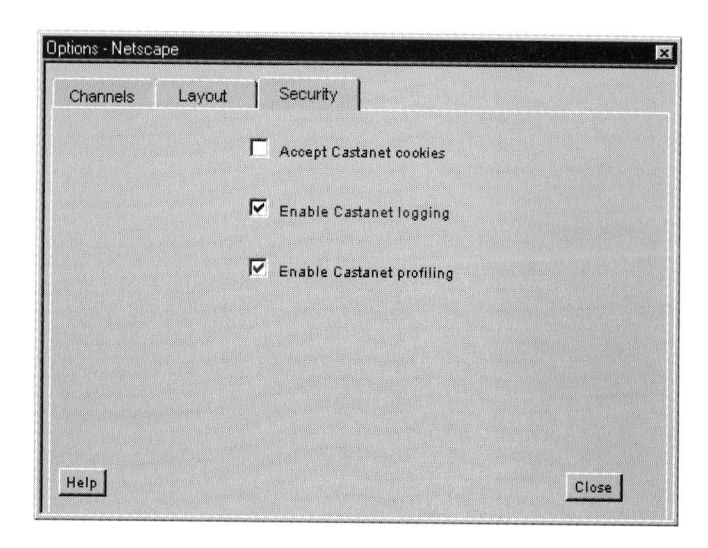

11. Check these options to enable Netcaster channels to send and receive information from your computer unencumbered. If you're uncomfortable allowing this access to your computer, uncheck the boxes. In most cases, you can leave them checked or leave the default settings in place (shown here).
12. Choose OK.

In this chapter, you learned how to automatically receive Web information using Netcaster. In the next chapters, you'll learn how to move beyond this one-way communication and join Internet-based communities using e-mail, discussion groups, and more.

CHAPTER

6

E-mail Made Easy: Using Netscape Messenger

INCLUDES

- Get your bearings in Netscape Messenger
- Set up your e-mail account
- Master the address book
- Set your e-mail preferences
- Create an e-mail message
- Get your e-mail

Start Netscape Messenger ➤ pp. 111-112

- From another Netscape Communicator 4.0 module, click the Mailbox icon on the Component Bar, *or*
- From the Windows 95 Start menu, choose Programs | Netscape Communicator | Netscape Messenger.

Retrieve Your E-mail Messages ➤ pp. 113-114

- Click Get Msg on the Netscape Messenger toolbar.
- Netscape Messenger will retrieve your e-mail messages.

Organize Your Incoming Mail ➤ pp. 114-115

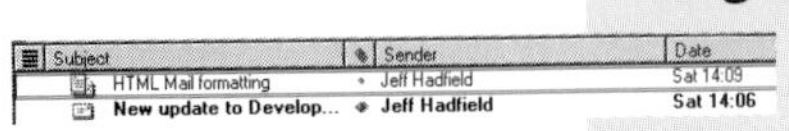

- Click the green diamond heading to list your unread messages first.
- Click the Date heading to list your newest messages first—or last.
- Click any other heading to sort messages by that heading.

Organize Your Addresses ➤ pp. 118-121

With Netscape Messenger's address book, you can keep track of often-used addresses.

1. Click New Card to create a new address book entry.
2. Fill in the blanks in the New Card dialog box, including name and e-mail address.
3. Click OK.

Create E-mail Messages ➤ *pp. 121-122*

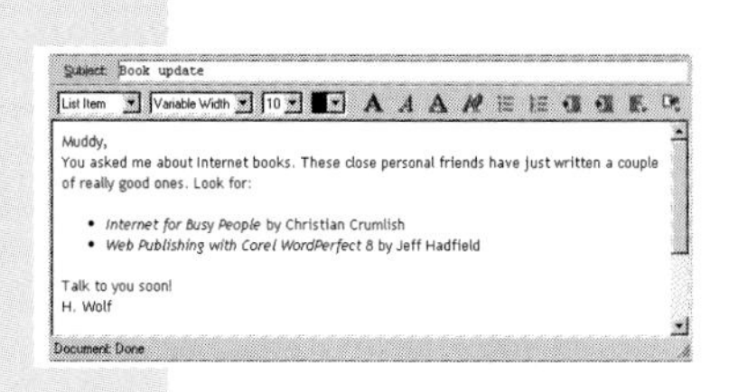

1. Click File | New | Message or click the New Msg button on the toolbar.
2. Type the e-mail address (or name, if they're in your address book) of the first recipient.
3. Attach any files or documents you want to send along.
4. Type a subject (make it descriptive).
5. Type your message.
6. Click the Send button or click File | Send.

Send Files and Other Stuff ➤ *pp. 121-126*

- Click the Attach button and hold the button down for a couple of seconds before you release it, then
- Choose File to attach a file to the e-mail message, then choose the file you want from the dialog box, *or*
- Choose Web page to attach a Web page, *or*
- Choose My address book card to attach your personal information.

Netscape Messenger is the key tool you'll use to send and receive e-mail messages. In this chapter, you'll learn about the things you'll do the most: create, send, receive, and organize your e-mail messages. You'll also get a glimpse of some of Netscape Messenger's other, more powerful abilities—but we won't waste your time showing you too much of the features you'll rarely use.

Is This for Me?

If you're going to use e-mail—and you probably will—you should read this chapter. We'll make it snappy, we promise.

The World Wide Web receives a lot of hype, and it's often thought of as the same thing as the Internet. Of course, you know it's *not*, but it's the most visual way to use it. Reality: e-mail is by far the most commonly used part of the Internet. For example, most of Jeff's family has e-mail—from his parents to his sister and sisters-in-law. On a recent trip to Seattle, he corresponded with a cousin about the trip before he even arrived.

You'll find e-mail indispensable. And Netscape Messenger will soon become indispensable, too. If you've used previous versions of Netscape's mail programs, you'll be impressed with the additional features. It's not an afterthought: Netscape Messenger is a full-featured, powerful e-mail program that will probably suffice for your e-mail needs.

EXPERT ADVICE

*For most people, Netscape Messenger's e-mail abilities are more than enough. But if you're looking for extremely powerful e-mail abilities, including being able to download e-mail from multiple servers and addresses in the same session, you might want to look at the justifiably popular Eudora Pro from Qualcomm (**http://www.qualcomm.com**). They also make a less powerful version—free for the download—called Eudora Lite.*

Starting Netscape Messenger

Enough discussion. Let's get started. But first, if you haven't set up your e-mail account, refer back to Chapter 2 to do so. We'll wait here until you get back.

EXPERT ADVICE

You can also refer to Chapter 2 to change things like the signature file that automatically appears at the end of messages you send or the e-mail address or name you use as the "reply to" address.

If you're already in another Netscape Communicator 4.0 module, like Netscape Navigator, you can launch Netscape Messenger by just clicking the Mailbox icon on the Component Bar:

SHORTCUT

If you launch Netscape Messenger from the Component Bar, it will automatically check for and download any new e-mail messages waiting for you on the server.

You can also launch Netscape Messenger from the Windows 95 Start menu: choose Programs | Netscape Communicator | Netscape Messenger.

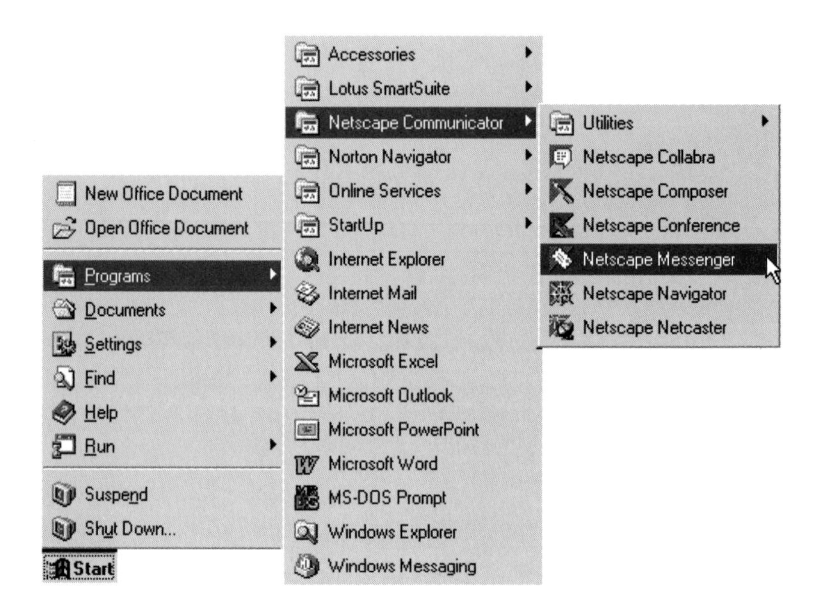

What's What in the Messenger Window?

Once you've started Netscape Messenger, you'll see the main window. When you start it for the first time, the list of e-mail messages will obviously be empty. But after you've received a few messages, it will appear more like the one in Figure 6.1.

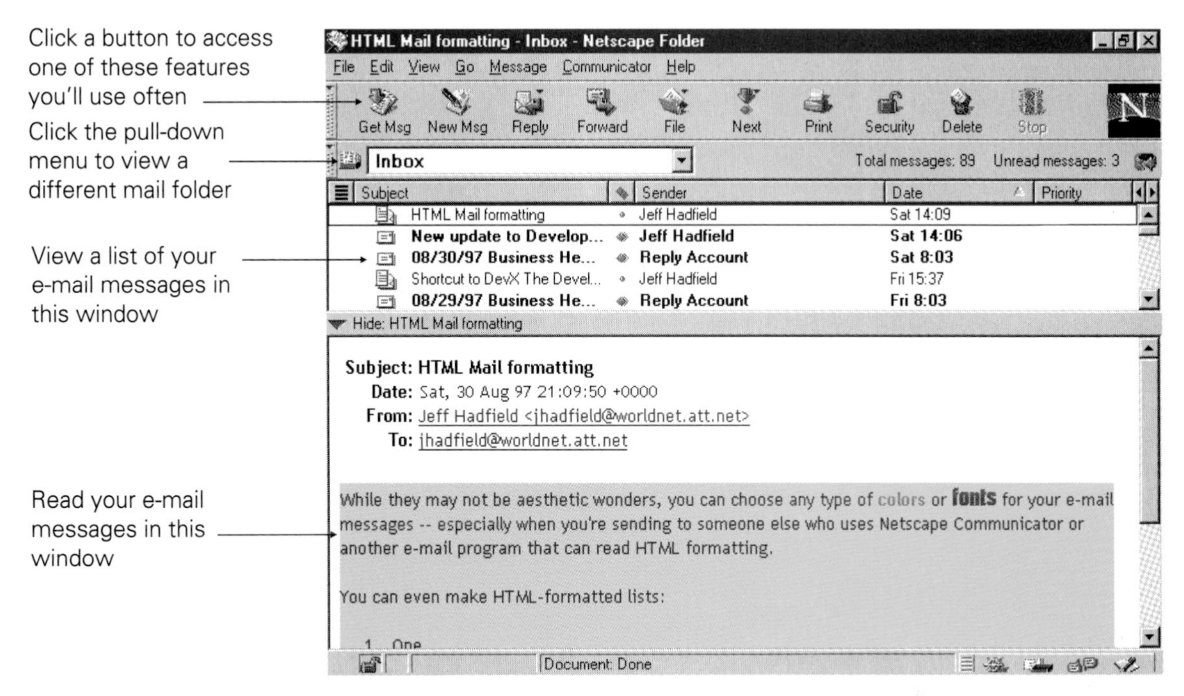

Figure 6.1 The main Netscape Messenger window

Quick Access

The toolbar—the menu of buttons at the top of the screen—lists the Netscape Messenger abilities you'll use a lot. You can access all of these from the pull-down menus, but you'll generally find it just as fast to use the buttons. Here's more info about each:

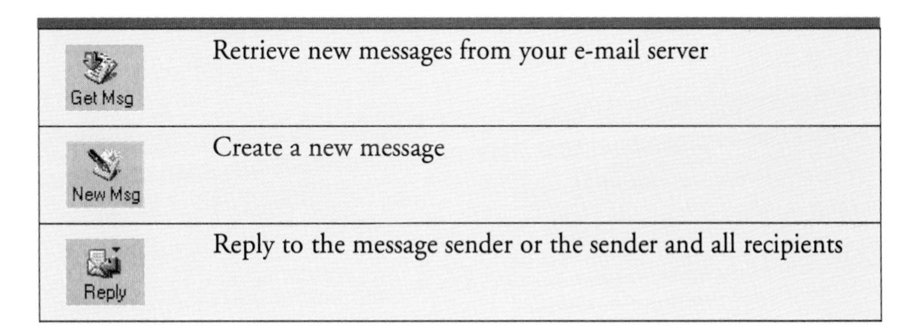

Get Msg	Retrieve new messages from your e-mail server
New Msg	Create a new message
Reply	Reply to the message sender or the sender and all recipients

Forward	Forward—send a copy of the message to someone
File	File the message in another folder
Next	Move to the next message
Print	Print the current message
Security	View security information about the message
Delete	Delete the current message
Stop	If there's something being downloaded or uploaded, stop the current information transfer

You may notice little green, downward-pointing triangles on three of the buttons: Reply, File, and Next. Click on any of the three and *hold* the button down briefly to get a menu of additional options. For example, the Reply button gives you the option to reply to either the message sender or the sender and all recipients. The File button gives you a list of all the folders in your e-mail box. And the Next button lets you move to the next unread message, the next thread, or several other options.

Viewing Lists of Your Messages

Netscape Messenger gives you a flexible way of viewing your incoming mail messages. You'll see a list of messages in the main window (Figure 6.1) that looks like the one below. Click on any of the headings to sort the list by that heading. The little triangle, or arrow, shows which heading the list is currently sorted by. When the arrow points up, the list is sorted in descending order; when it's pointing down, the list is sorted in ascending order (first to last or newest to oldest).

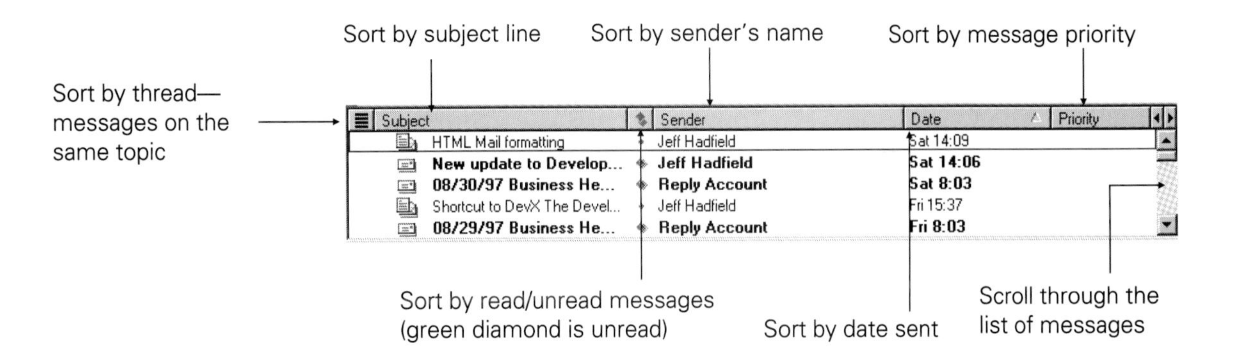

In the list, unread messages are shown with a small, closed envelope icon to the left of the subject line. Once you've opened the message, the icon changes to resemble an opened letter. Unread messages are listed in bold; read messages are in regular text.

Reading Messages

To save you the effort of opening each message to read it, you can use part of your screen to view the message (again, take a look at Figure 6.1 to see how). Click *once* on any message listed in the message list window to view it in the preview window. Click twice to open the message in a separate window (Figure 6.2).

The preview window, however, displays messages just as they would appear in a separate window. Viewing messages is pretty straightforward, but here are a few special tips:

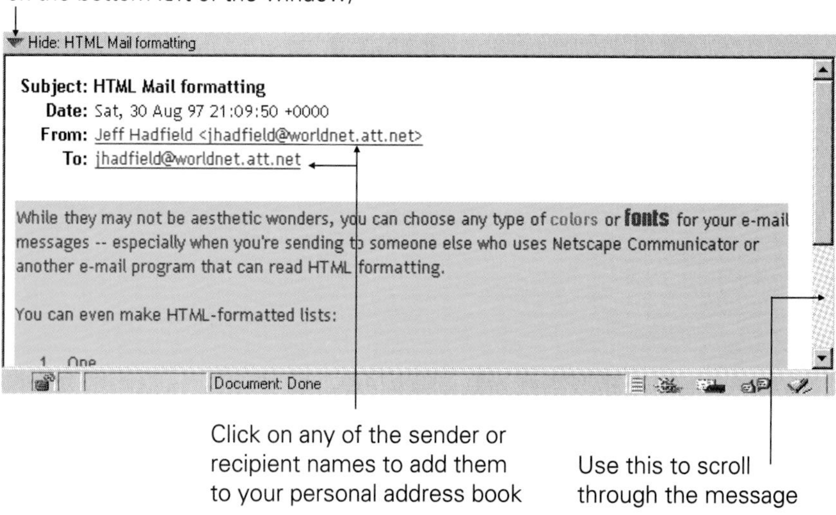

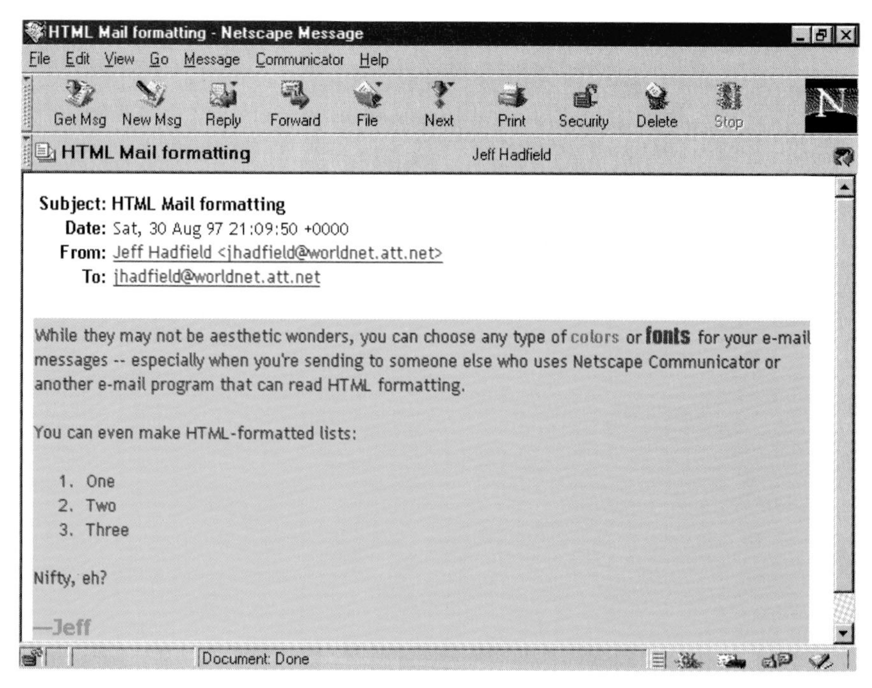

Figure 6.2 If you want to see more of your e-mail message, you can view it in a separate window like this one

EXPERT ADVICE

Sometimes you'll find that an e-mail message someone sent you is difficult to read because there are no line breaks in a paragraph. Instead of using the horizontal scroll bar to scroll through the message, just choose View | Wrap Long Lines. The message will be reformatted to fit inside the window.

Save Typing with Your Own Address Book

Like a personal address book you might carry in a bag or briefcase, Netscape Messenger lets you create your own electronic address book for commonly used addresses. The address book can contain more than just names and e-mail addresses—although, realistically, those are the only two bits of information you'll regularly use. However, you can use the address book as a lightweight contact manager if you choose.

You learned above how to click on a name in an incoming e-mail message to add it to your address book. Next, you'll learn how to access the address book when you create an e-mail message.

Before you do, though, take a quick look at the address book itself. To view Netscape Messenger's address book, click Communicator | Address Book. You'll see the window shown in Figure 6.3.

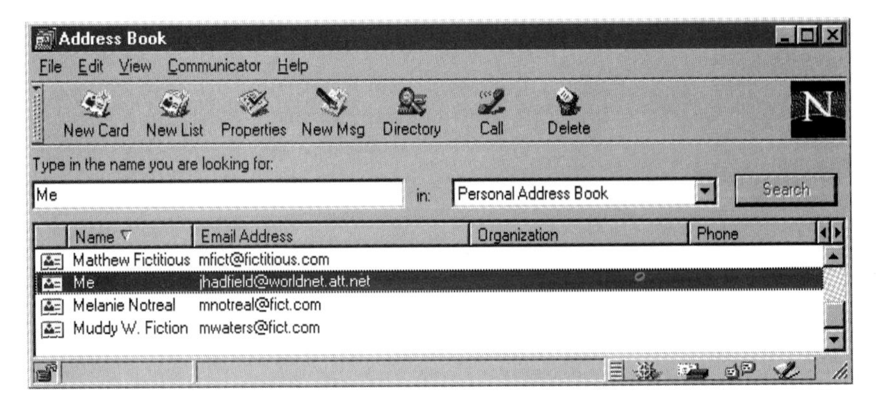

Figure 6.3 Netscape Messenger's Address Book window. Manage your e-mail contacts with this utility

DEFINITIONS

Address card: *Roughly corresponds to a Rolodex card. You can create a card for any of your e-mail contacts.*

List: *Like a mailing list—several people to whom you regularly mail something as a group. A list is made up of several e-mail addresses that exist as cards.*

You'd think an address book program would be as simple as a Rolodex to use, right? Well, it's not quite that simple, but it's close. Here's a field guide to identifying the Address Book's icons in the wild:

New Card	Create a new contact card
New List	Create a new mailing list
Properties	Change the currently highlighted card's information
New Msg	Create a new message addressed to the currently highlighted person or list
Directory	Look up someone's information in an online directory (see Chapter 4)
Call	Call someone using Netscape Conference (see Chapter 8)
Delete	Delete the currently highlighted contact card

Adding Someone to Your Address Book

Most often, the only thing you'll really do with your address book is add people to it. No matter how you choose to add people—whether it's from the address book itself or from within an e-mail message—you'll end up filling in the same set of information.

EXPERT ADVICE

Although we won't delve into it here—remember, we're trying to show you the basics, fast—if you have an existing electronic address book, Netscape Messenger can import your data if it's in a common format like ASCII. If you want to play with it, go ahead, but Jeff finds it easier to copy and paste from his contact manager for the few addresses he doesn't use often.

To add someone to your Address Book, you'll need to follow a few simple steps. Here we go:

1. From the Address Book window, click the New Card button (or choose File | New Card).
2. Under the Name tab, fill in the first name, last name, organization, title, e-mail address, and other information. The first name, last name, and e-mail address are essential.
3. Check the "Prefers to receive rich text (HTML) mail" box if you know the person can receive HTML-formatted mail (especially if you know they're using Netscape Communicator 4.0).

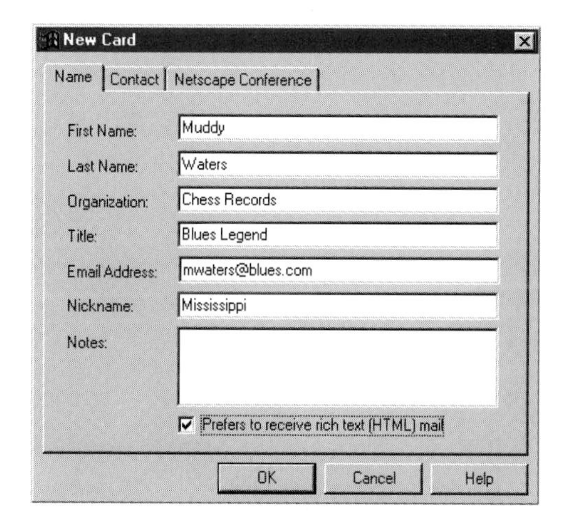

4. Click the Contact tab to add more information, *or*

5. Click OK to save the card and move on.
6. Fill in any of the address information you want to include.
7. Add business and home phone numbers and a fax number, if you like.

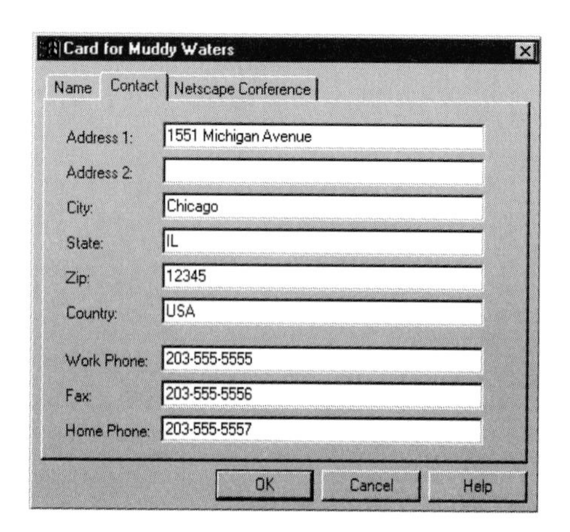

8. Click the Netscape Conference tab if you want to change those settings, *or*
9. Click OK to save the card and move on.
10. If you know the recipient's Netscape Conference server and want to change it from the default Netscape Conference server, choose Specific DLS Server or Hostname or IP Address from the pull-down menu.
11. Type the DLS name, hostname, or IP address in the text box.

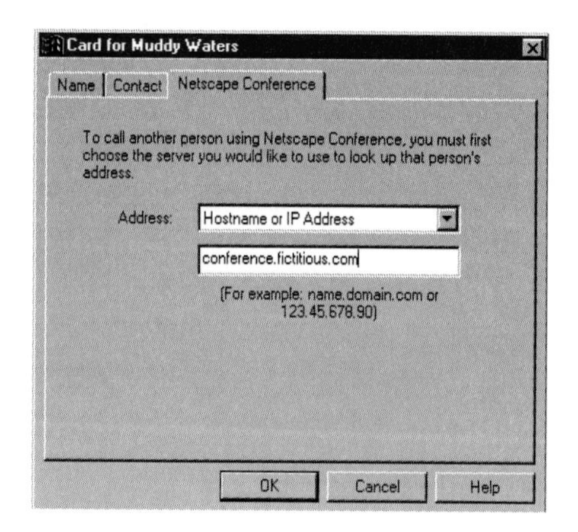

12. Choose OK to save the card and close the window.

EXPERT ADVICE

Don't worry too much about changing the Netscape Conference Server. In most cases, someone will have provided you with the correct conference information. Otherwise, you can just leave it alone.

Creating E-mail Messages

What's the point of e-mail if you can't create messages? Well, of course you can in Netscape Messenger. And you can do it with a variety of formatting tools. The basics of creating e-mail messages are easy. Take a look at those—then we'll talk for a while about a few advanced capabilities that can save you some time and money—as well as some tips that can save you electronic embarrassment.

Follow the Step by Step instructions on the next page. That's really all there is to sending e-mail messages. For most of the messages you'll send, those are the only steps you'll need. Notice in the Step by Step how part of the address is shown in gray: as you begin to type someone's name, if you have that person listed in your address book, Netscape Messenger will automatically complete the name for you. If the name is correct, you can just press ENTER to move on.

EXPERT ADVICE

You can attach just about any kind of file to any e-mail message. However, out of respect for other people's time and money, you should keep attached documents as small as possible. You can compress documents using shareware programs such as PKZip and WinZip for Windows and Stuffit for the Mac.

Some Fancy E-mail Stuff

As with regular mail, e-mail can do several fancy things beyond just sending a basic message. You can send the message to carbon copy recipients for informa-

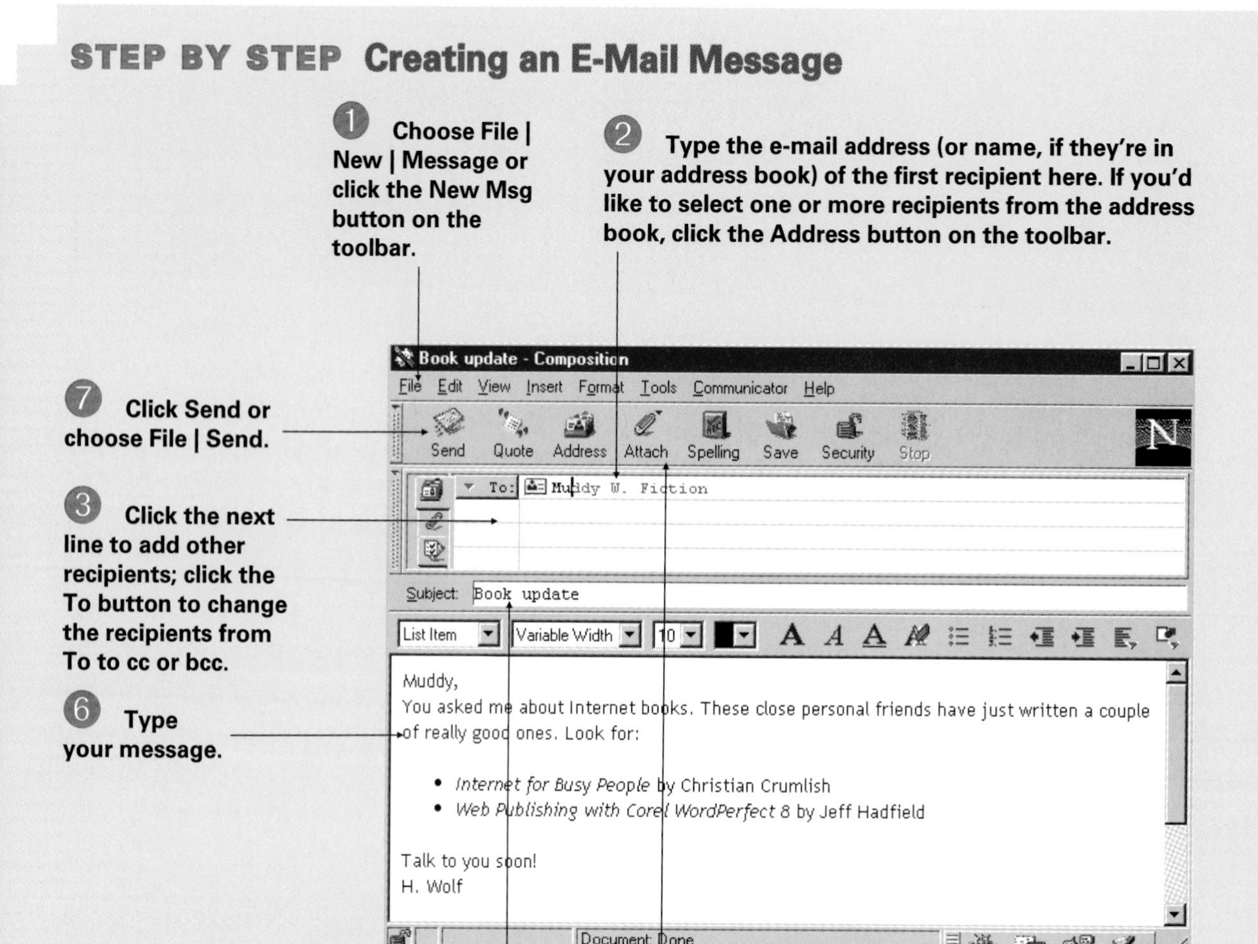

tion purposes—or even to blind copy recipients, where the other recipients don't see that person's name. You can attach multiple files (as shown in Figure 6.4), request a receipt, quote previous messages, and save the message in a draft form.

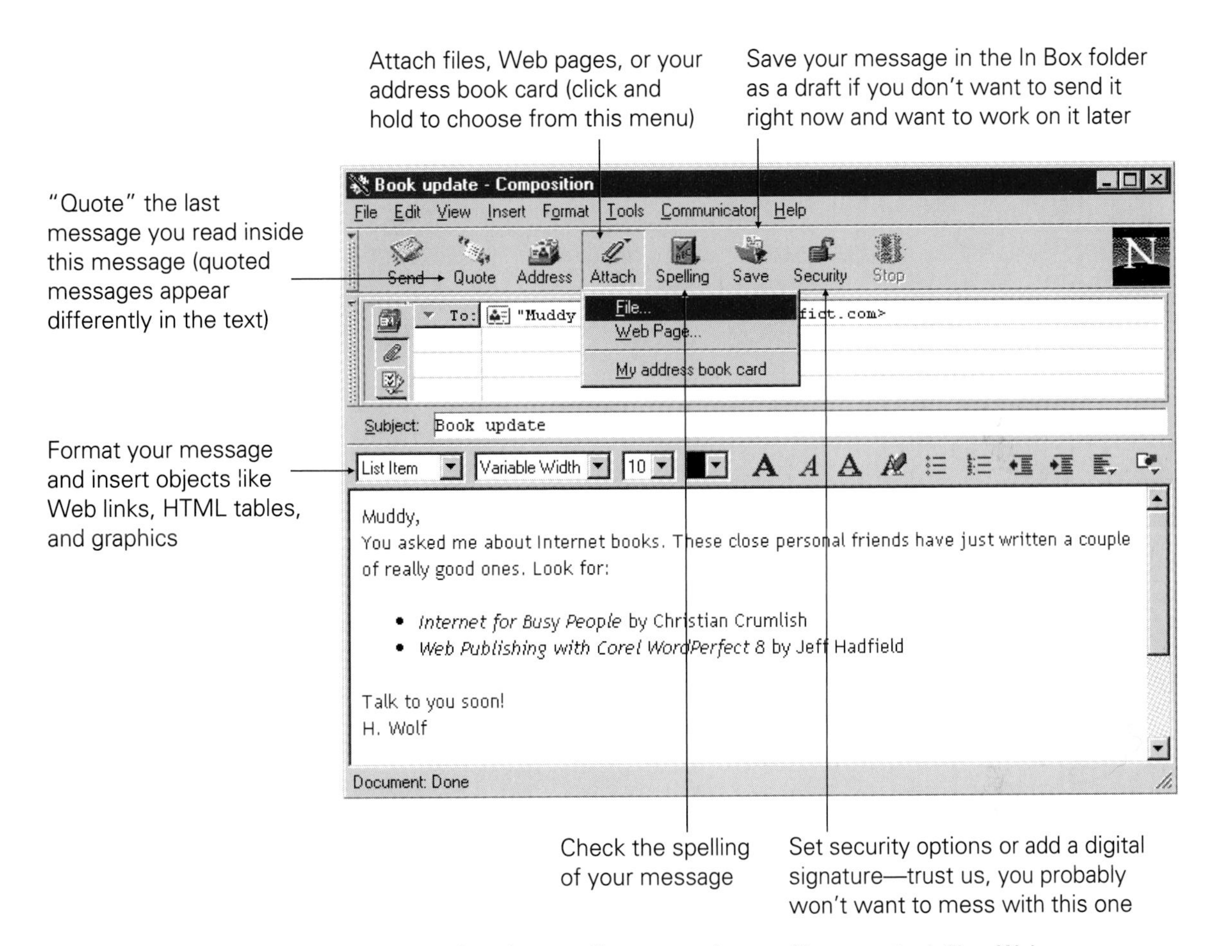

Figure 6.4 Creating e-mail messages is easy. You can attach files, Web pages, or even your address book card

You may want to adjust some additional settings. To view most of the message's advanced settings, click the settings tab on the left side of the address list (below the address tab and the attachments tab). You'll see these settings:

Encrypted
Signed
Uuencode instead of MIME for attachments
Return Receipt
Priority: Normal
Format: Ask me

Most of the time, you will never use the three settings on the left— they can help you send secure information or cope with incompatible file attachments. The three settings on the right are more useful in day-to-day communication.

If you'd like to receive an automated message from the recipient's e-mail system that the message has been received, check the Return Receipt box. If you want the message to arrive with a special priority—usually either *high* or *low*, choose it from the Priority pop-up menu. When they arrive, high-priority messages appear in most e-mail systems with a red icon of some kind (like an exclamation point). Low-priority messages usually appear with a different icon, like a blue, downward-pointing arrow.

Finally, if you want to specify the format in which to send the e-mail message, specify it here. By format, Netscape Messenger means either plain text—with no formatting whatsoever—or HTML format, with all the formatting tools you can use on a regular HTML Web page intact. If you *don't* specify a format here, Netscape Messenger will ask you when you send the message what format you want to send it in (unless you've marked the recipients in the address book as being able to receive HTML-format mail). Our opinion? For most e-mail messages, you can pretty much stick with plain text. If formatting makes the meaning of your message more clear—and isn't included just to gussy it up—then you might want to send it in *both* plain text and HTML format. If the recipient can't read HTML formatting, at least the plain text will be intact.

Avoiding Embarrassment

Entire books have been written on basic e-mail etiquette. But we'll tell you a few key items that will spare you from looking like a rube—or, worse, getting "flamed" for unknowingly breaking some Net custom.

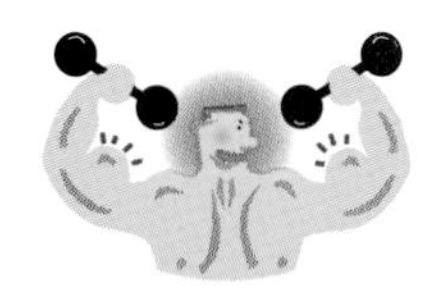

DEFINITION

Flaming: *Sending a scathing, often rude diatribe to someone with whom you disagree or whom you feel has transgressed propriety in some manner. It's the Net equivalent of yelling at someone—but many flamers will say things in text they'd never say to someone face-to-face.*

Here's a handful of key rules you'll be glad you know:

- **Provide a meaningful subject line.** "Hi" doesn't help anyone see the value in reading your e-mail message. Instead, it should be a summary of the message's contents.
- **Quote to provide context.** You don't always have to include an entire previous message in your replies. Instead, quote enough of the original message to show what you're replying to so the recipient won't take the information out of context.
- **Don't use all caps.** Use a mixture of upper and lowercase letters, just as you would normally. USING ALL CAPS MAKES YOUR MESSAGES SEEM LIKE A TELEGRAM—OR, WORSE, LIKE SHOUTING.
- **Don't forward your e-mail messages** to anyone, especially a public discussion group, without the permission of the sender.
- **Informality is okay.** We all learned the rules of formal correspondence in school, but e-mail more closely resembles conversation.
- **Learn e-mail acronyms and smileys to convey your message.** Like anything, a little of these goes a long way—but when you're attempting humor in your e-mail messages, using a smiley like this :-) or <g> to mean "grin" can help ensure that your message is taken in the right light.

Web sites that can help you learn more about e-mail etiquette, acronyms, and smileys are listed in Appendix A.

Built-in Memory

Ever wish you could remember exactly what you'd written or said in a conversation? Netscape Messenger helps you do just that. Every message you send is automatically saved in a Sent Messages folder. So if you're wondering what you said, just choose the Sent folder from the folder pop-up menu.

EXPERT ADVICE

After using Messenger for a while, your Sent folder can get pretty full and will take up valuable disk space. It's probably a good idea to clean up this folder once in a while by deleting old, unwanted messages. Highlight all the messages in the Sent folder and drag them to the Trash folder—or just press Delete. Then, choose File | Empty Trash Folder to delete them permanently from your hard drive (otherwise, they'll just sit there in the trash can).

Working on E-mail When You're Not Connected

Netscape Messenger and Netscape Communicator include a lot of neat features for working on the Internet even when you're not connected to it. Realistically, however, you'll find they're often more confusing than helpful.

Is This for Me?

If you're using Netscape Communicator 4.0 and Netscape Messenger while connected to the Internet all the time—at work, for instance—you'll probably never use these features. If you're using a dial-up connection—connecting over a phone line—you might.

The easiest tip to try: launch Netscape Messenger and download your e-mail messages (click Get Msg). Once all the messages are retrieved, disconnect from the network. You don't need to be connected to the Internet while you're reading your e-mail, since your e-mail is downloaded to your hard drive.

In a similar way, you can create your e-mail messages, save them to your hard drive, and then send them all at once when you're connected to the Net. Just create your e-mail messages normally—but instead of choosing Send, choose File | Send Later. Once you've connected from the main Netscape Messenger window, choose File | Send Unsent Messages. Netscape Messenger will send your e-mail messages in one big batch.

Working on your e-mail when you're not connected makes it easy to work on, say, your laptop when you're traveling and not connected to a phone line. Or it can help keep you from tying up your home line when you're reading your e-mail from your home computer.

EXPERT ADVICE

You may also want to explore the online/offline options Netscape Messenger offers. These can upload and download e-mail and discussion group messages each time you log on or log off.

Getting E-mail

If you've ever used an office e-mail system, you may be accustomed to your e-mail automatically arriving in your mailbox. With a dial-up Internet connection, that's not the way it works. Instead—and this is a feature, not an inconvenience—you just need to ask Netscape Messenger to check for new messages.

To check for new messages, click Get Msg on the toolbar or choose File | Get Messages | New. If you have any messages, Netscape Messenger will retrieve them from your e-mail server. It'll show you the progress as it downloads like this:

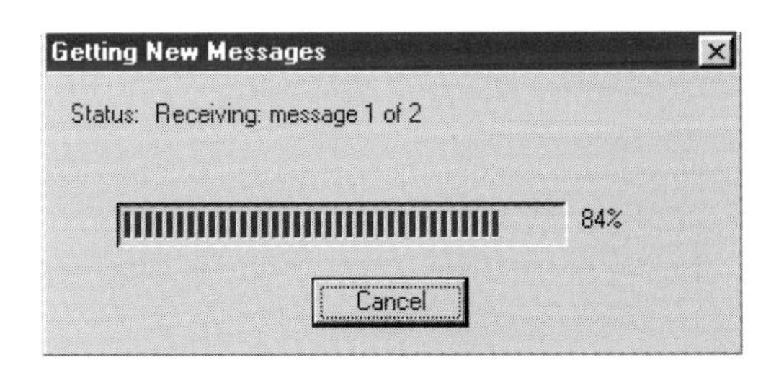

Once you've made it this far, you've mastered the basics of sending and receiving e-mail. In the next chapters, you'll learn about other ways to communicate with others, including real-time collaboration and Internet newsgroups.

CHAPTER

7

Group Discussions: Newsgroups and More

INCLUDES

- Finding a list of newsgroups
- Finding the newsgroup you want
- Subscribing to a newsgroup
- Reading a newsgroup
- Posting a message in a newsgroup

Start Netscape Messenger ➤ pp. 132-133

- From another Netscape Communicator 4.0 module, click the Mailbox icon on the Component Bar, *or*
- From the Windows 95 Start menu, click Programs | Netscape Communicator | Netscape Messenger.

Get a List of Newsgroups ➤ pp. 134-136

1. Choose File | Subscribe to Newsgroups.
2. Click Get Groups.

Search for Newsgroups ➤ pp. 136-138

1. If you haven't already, launch Netscape Messenger and then choose File | Subscribe to Discussion Groups.
2. Click the Search for a Group tab.
3. Type the word you want to look for.
4. Click Search Now.
5. From the list of results, highlight the newsgroup you want to subscribe to.
6. Choose Subscribe to add it to your list of newsgroups.

Read Newsgroups ➤ pp. 139-143

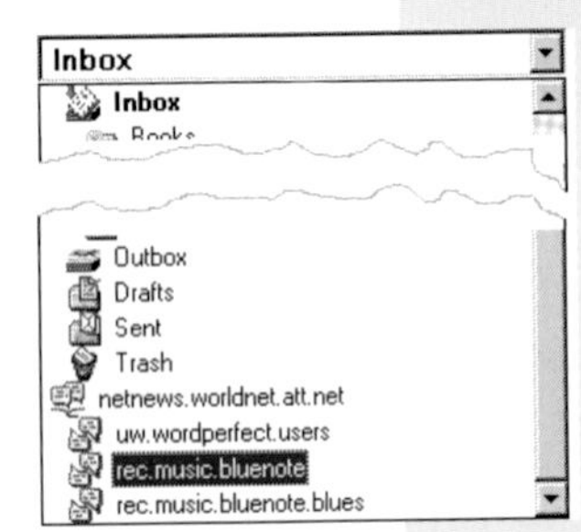

1. From the drop-down menu, choose the newsgroup you want from your list of subscribed newsgroups.
2. Scroll through the message topics.
3. Click on the message you want to read.

Reply to Newsgroup Messages ➤ *pp. 141-142*

1. Highlight the message you want to reply to.
2. Click the Reply button on the toolbar.
3. Type your reply.
4. Click Send.

Create and Reply Without Embarrassment ➤ *pp. 143-144*

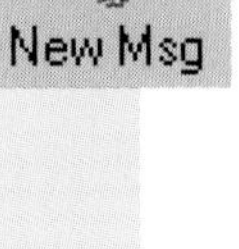

Newsgroup postings are like e-mail, but they have their own quirks. Remember to

- Read the FAQ—a document of frequently asked questions.
- Hang around in a newsgroup first to make sure you understand the "culture" before you post.
- Practice in a practice newsgroup first.
- Look in an "answers" newsgroup for help.

In the last chapter, you learned how to use Netscape Messenger for e-mail—but you can do more than just that. You can also use it to participate in group discussions with other people around the world. These group discussions are sometimes called USENET ("Users' Network")—a vestige of the "old" Internet—but are also called Netnews or newsgroups.

Newsgroups form the backbone of the Internet—they were popular long before the Web was—and are centered around the Internet's ability to bring people of like interests together. No matter where they live, groups of people can discuss things they're all interested in—from esoteric computer programming to childcare.

Is This for Me?

Well, Internet newsgroups can contain valuable information. But they can also be an easy way to waste a lot of time—both because they're notorious for containing information that's off the topic and because they can compel you into habitual use. You may find searching for information on the Web more useful—or using a search engine like DejaNews (**www.dejanews.com**) to search newsgroup postings.

Netscape Messenger makes it easy to find newsgroups in which you're interested—and to keep up to date on them and participate in them as well. To do so, you'll first need to launch Netscape Messenger.

If you're already in another Netscape Communicator 4.0 module like Netscape Navigator, just click the Mailbox icon on the Component Bar.

SHORTCUT

If you launch Netscape Messenger from the Component Bar, it will automatically check for and download any new e-mail messages waiting for you on the server.

You can also launch Netscape Messenger from the Windows 95 Start menu: just choose Programs | Netscape Communicator | Netscape Messenger.

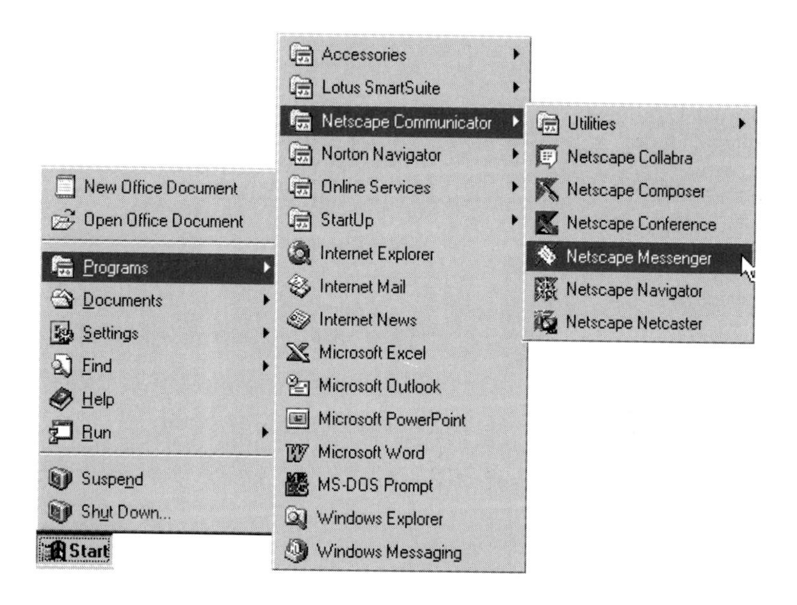

Once you've launched Netscape Messenger, you'll need to choose some newsgroups to get started.

CAUTION

Be sure to connect to the Net while you're doing the things listed in this chapter. Until you've mastered working online and offline, you're better off staying connected when accessing newsgroups. It's hard to predict when Netscape Messenger will need to access the Net, especially when setting things up.

Getting Started with Newsgroups

If you haven't already set up Netscape Communicator 4.0 with your news server information, take a moment to look back at Chapter 2 and do so.

Not surprisingly, Netscape Communicator 4.0 can't read your mind. You'll need to help it a little. You can do this by telling it what newsgroups you want it to keep track of for you.

You'll first need to retrieve a list of newsgroups currently available. Choose File | Subscribe to Discussion Groups. You'll see the window shown in Figure 7.1.

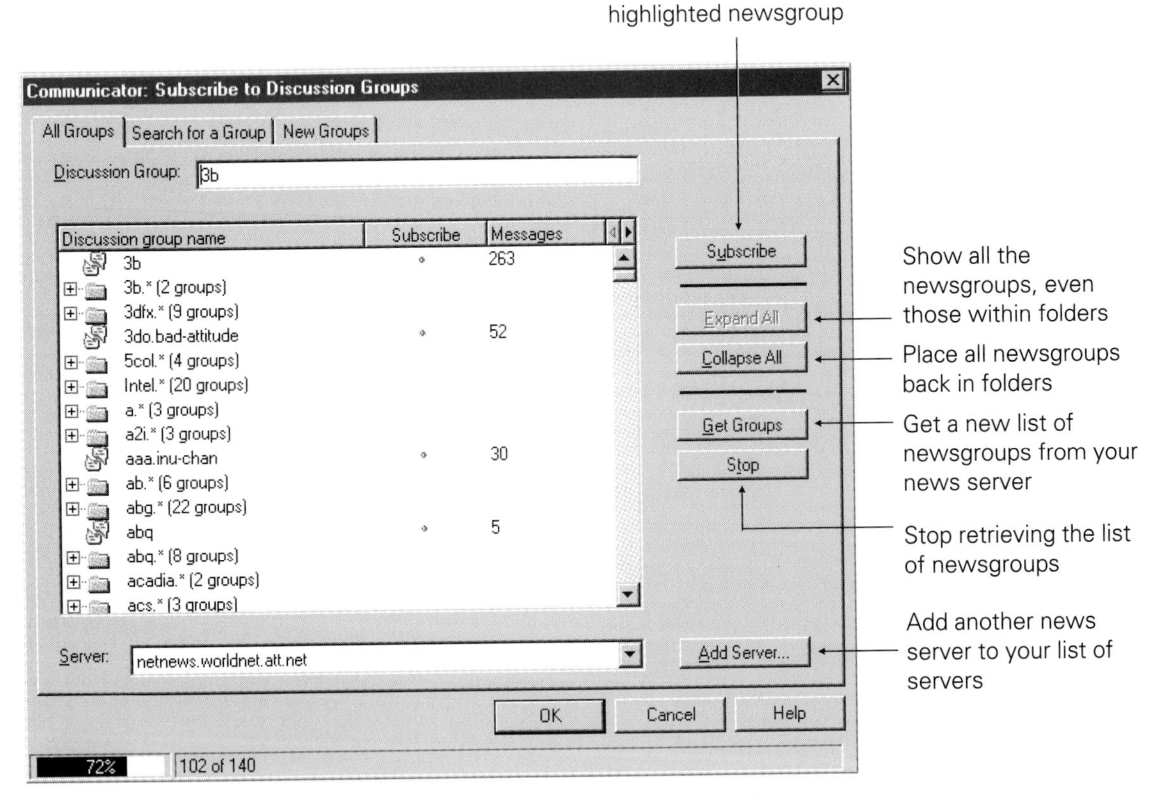

Figure 7.1 Use this window to get a list of discussion groups—and subscribe to them.

Getting a List of Newsgroups

When you set up your news options in Netscape Communicator 4.0, you specified a main news server. For the most part, you'll use that server for all your news—it includes all the generally available, Internet-wide newsgroups.

Get Groups

But, unlike Figure 7.1, your list of newsgroups will be empty when you first run Netscape Messenger. Fill your list by clicking Get Groups.

EXPERT ADVICE

The Net contains thousands of newsgroups. Odds are, even on a fairly fast connection, it'll take a while to download the entire list. Be patient. And if you can't wait any longer, click Stop.

Once you've downloaded the entire list, your list will appear like the one in Figure 7.1. They'll be listed alphabetically. Groups on similar subjects are grouped into folders. Click on the plus sign to the left of the folder to display the contents of the folder, like this:

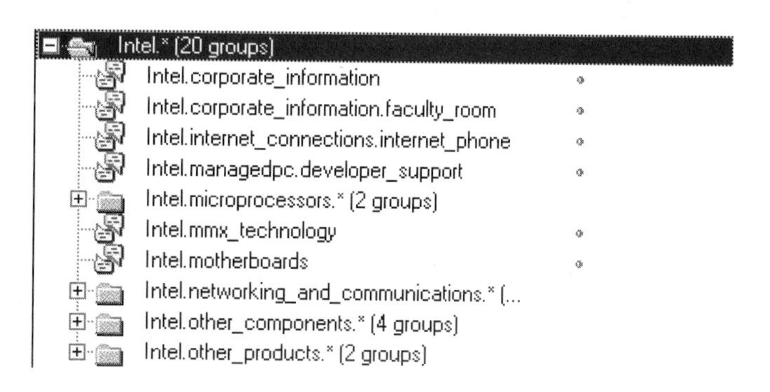

Once you've got this monstrous list, you can browse through it to find the information you need. Table 7.1 lists the most widely used types of newsgroups, although others exist (like the popular "alt" alternative newsgroups). Newsgroup names will start with one of the words in this table followed by a period and then more identifiers, like this: *rec.autos.vw.*

Newsgroup name starts with	Newsgroup is about
comp	Computers
misc	Miscellaneous topics
news	News about newsgroups and Usenet (not general, CNN-style news)
rec	Recreational topics like sports, music, and travel
soc	Social or societal topics
sci	Science-related topics
talk	Talk and debate (much like talk radio)

Table 7.1 The Official "Big Seven" Newsgroup Types

Once you've gotten in the habit of looking at new newsgroups, you can use a nifty Netscape Messenger feature to keep track of them for you by clicking the New Groups tab, as shown in Figure 7.2. Choose Clear New to erase the current contents of the list, if any. Next time you log on, the list will contain newsgroups that have been added since the last time you checked. To subscribe to any of these newsgroups, select it and click Subscribe.

Finding a Particular Newsgroup

If you don't want to slog through the entire list of Internet newsgroups, you can just search for the one you want.

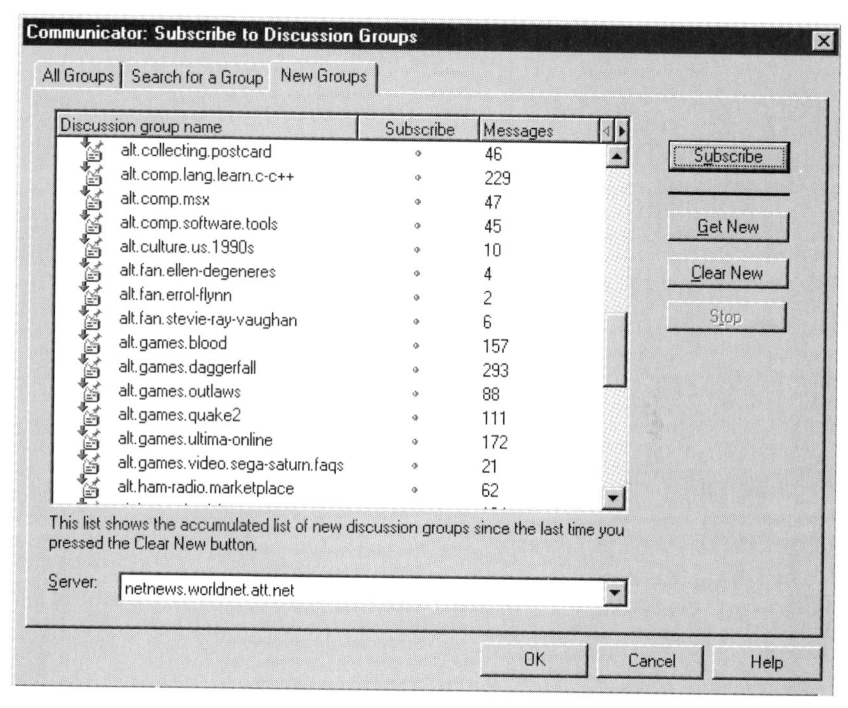

Figure 7.2 Use Netscape Messenger's new newsgroup tracking to watch for recent additions

CAUTION

Searching for newsgroups is far from sophisticated: the words you type must be found in the newsgroup name exactly as you've typed them. For example, if you're looking for discussions about Volkswagens, you wouldn't find ***rec.autos.vw*** *if you searched for "Volkswagen"—you have to search for "vw."*

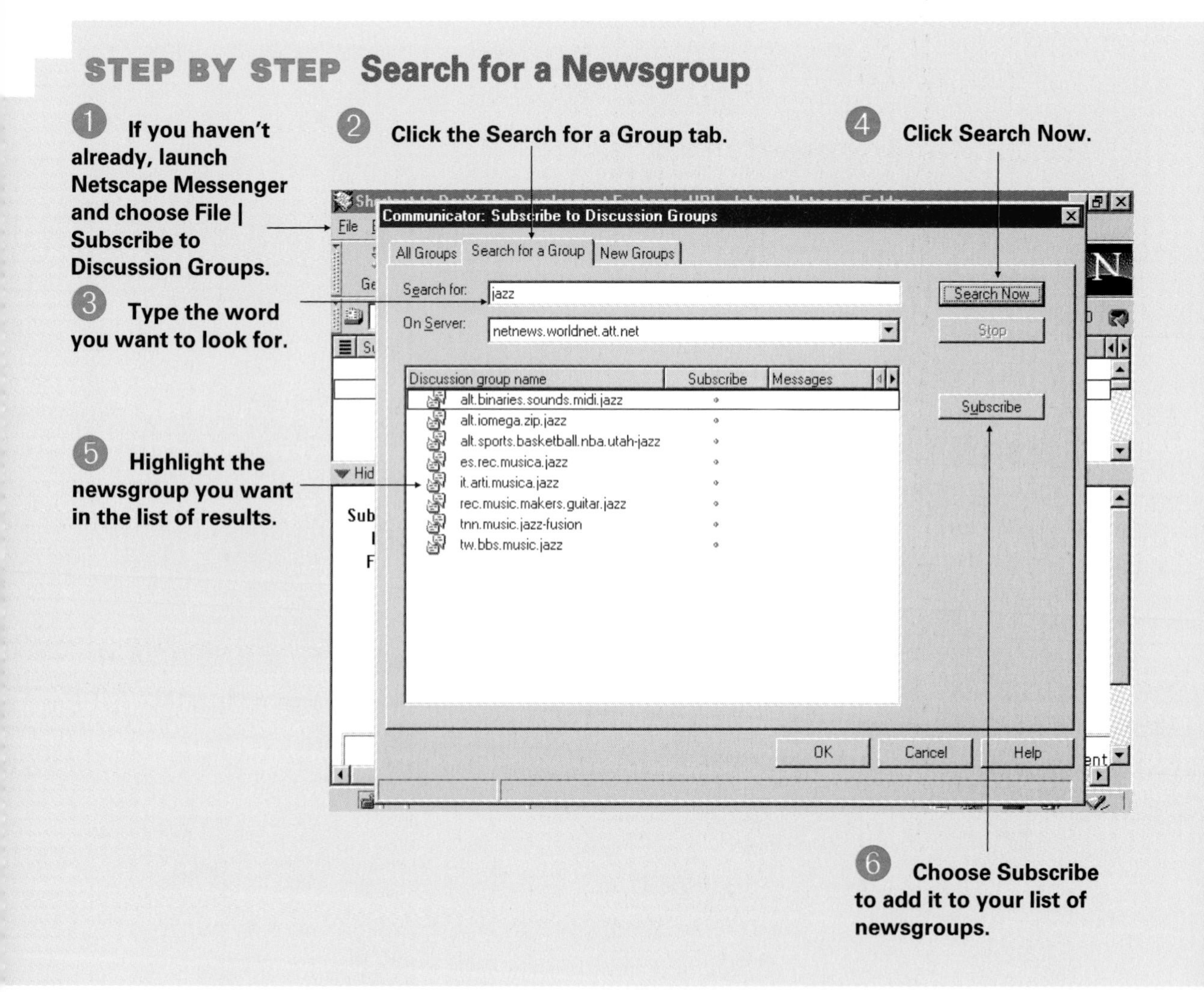

Subscribing to Newsgroups

To become a part of one of these discussions, even if you're planning only to listen, you need to subscribe to it. Fortunately, it's easy to subscribe (and unsubscribe). On any of the three tabs you've just looked at in the Subscribe to Discussion Groups window, just highlight the newsgroup you like, then click Subscribe.

Newsgroups to which you've subscribed appear with a checkmark, like this:

Discussion group name	Subscribe	Messages
rec.music.bluenote	✓	
rec.music.bluenote.blues	✓	

EXPERT ADVICE

After you've chosen the newsgroups you want, you can close the Subscribe to Discussion Groups window at any time by choosing OK.

Reading Newsgroup Messages

Great! You've done it. You've subscribed to a newsgroup or two. Now what? Well, it's time to read them. Luckily, if you read Chapter 6, you pretty much already know how to read newsgroup messages: you read them like you read e-mail messages. We'll show you the few, minor differences.

First, choose the newsgroup you want to read from the folder pull-down menu.

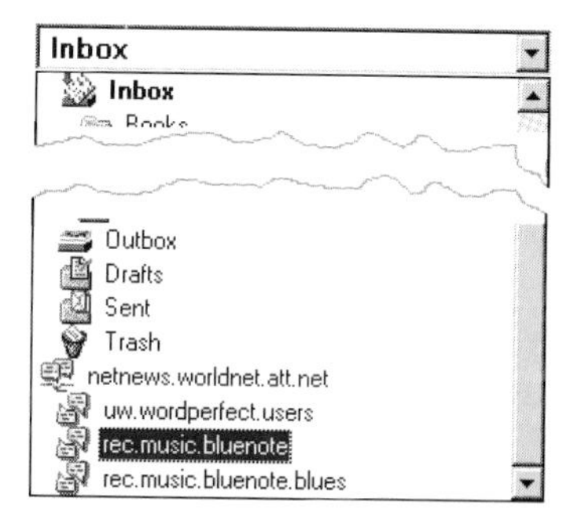

Since many newsgroups contain thousands of messages, Netscape Messenger will display the dialog box shown in Figure 7.3, asking your permission before it downloads all the message headers. You may want to download only a couple hundred headers, then download more when you reach the end of the list.

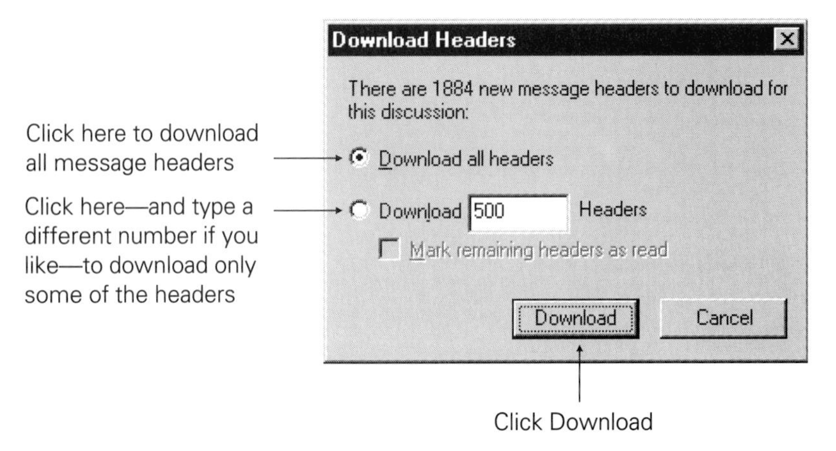

Figure 7.3 Choose the number of message headers you want to download initially when you load a newsgroup folder

DEFINITION

***Message header:** Contains information that helps you determine what the message contains: the sender's name, the subject line, the date and time sent, and so on.*

Netscape Messenger will show you its progress in downloading the message headers at the bottom of the screen. After it's done, you'll see a list of headers—with the highlighted message shown in the preview window—just like you do when reading e-mail. Take a look at Figure 7.4 to get your bearings in the Netscape Messenger window. Table 7.2 describes how each icon on the toolbar can help you manipulate your messages.

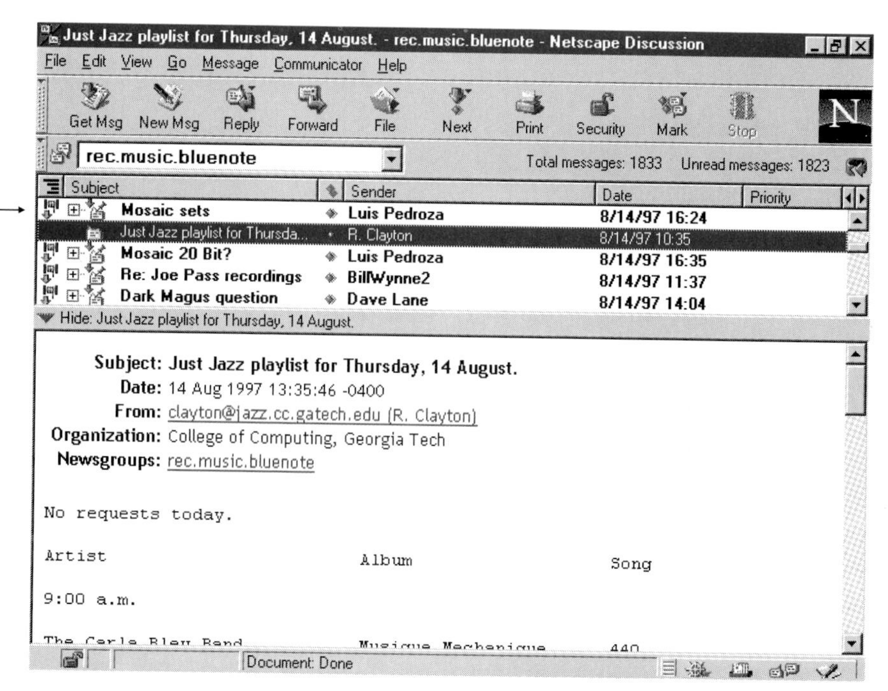

Figure 7.4 Read your newsgroup messages just like your e-mail messages

Reading, replying to, and forwarding newsgroup messages work just the same as e-mail (see Chapter 6). But there's one button here that's not on your e-mail screen: the Mark button.

Since Netscape Messenger doesn't download messages to your hard drive unless you specify, you can mark certain messages as read or unread. When you mark a message or message thread as read, it won't appear the next time you read the newsgroup messages—so you don't have to keep downloading messages you've already read.

DEFINITION

Message thread: *A string of messages on a related topic—replies and replies to replies. Netscape Messenger marks threads with a little spool icon to the left of the message line.*

Get Msg	Get more message headers
New Msg	Create a new newsgroup message
Reply	Reply to the currently highlighted message
Forward	Forward the currently highlighted message
File	File the currently highlighted message in one of your personal message folders
Next	Read the next message (click and hold for other options)
Print	Print the current message
Security	View security information about the current message
Mark	Mark the current message as read, unread, or for later reading (click and hold for options)
Stop	Stop the current download

Table 7.2 Icons for Manipulating Newsgroup Messages

Usually, you'll just skim a newsgroup for postings of interest, read them (they'll be automatically marked as read once you've clicked on them), and then come back another day. But you'll end up with a list of all the messages you didn't want to read the last time you checked in if you don't mark all the messages as *read* before you move on to the next newsgroup. How do you do that? Just click and hold the Mark button for a second or two, then choose All Read from the menu.

EXPERT ADVICE

You'll find some of the other mark functions useful as well. You may want to spend a few minutes experimenting with them—especially Mark for Later, which saves messages so you can read them after you've disconnected from the Net.

Must-Know Newsgroup Tips

Netscape Messenger's easy-to-use, combined interface for both e-mail and newsgroups is a bit deceiving. The two types of communication are similar, but you'll need to watch out for a few things when you're participating in newsgroups.

Read the FAQ Just about every newsgroup has a FAQ—Frequently Asked Questions—file or message that's regularly announced or posted. Read it before you ask a question that everyone's already answered hundreds of times.

Lurk First Lurking—listening to newsgroup conversation without participating—is valuable. It can prepare you to understand the newsgroup's culture, conventions, and even courtesies.

Practice or Look for Answers If you're unfamiliar with newsgroups and want to practice, try one of the many practice newsgroups available. You may also want to look for one of the "answers" groups—like **news.answers**.

EXPERT ADVICE

From Netscape Navigator, there's a quick way to check on any newsgroup without subscribing. In the Go to box, type news: *followed by the newsgroup name, like this:* ***news:rec.answers***.

While there are many things Usenetters can discover that we haven't covered here, you've mastered the basics of newsgroups in only a few minutes. After reading this chapter, you're ready to jump in and explore the vast amounts of (occasionally useful) information Internet newsgroups contain. In the next chapter, we'll take a quick, flyby look at some of the less-often-used collaboration tools included in Netscape Communicator 4.0.

CHAPTER

8

Collaborating and Chatting Over the Net

INCLUDES

- Launching Netscape Conference
- Starting a conference call
- Using a shared whiteboard
- Browsing the Web together
- Exchanging files
- Chatting using text
- Sending Internet voice mail

Launch Netscape Conference ➤ p. 149

- From the Windows Start Menu, choose Start | Programs | Netscape Communicator | Netscape Conference, *or*
- If you're already running another Netscape Communicator 4.0 module, choose Communicator | Conference.

Start a Conference Call ➤ pp. 154-156

1. If you don't know someone's e-mail address offhand, look it up in the Web-based directory or your own address book.
2. Type the e-mail address of the person you want to call.
3. Click the Dial button.

Use a Shared Whiteboard ➤ pp. 156-158

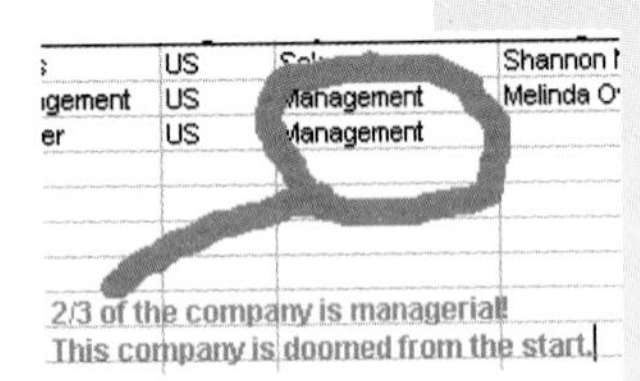

1. Launch the shared whiteboard by clicking the whiteboard button on the toolbar or by choosing Communicator | Whiteboard.
2. Make sure the screen or document you want to capture appears behind your whiteboard window.
3. Choose Capture | Desktop (or Window or Region if you don't want the whole desktop).
4. When you're returned to the whiteboard window, position the crosshairs in the window where you want the upper left corner of your screen image to lie.
5. Click once to place the image in the whiteboard screen.
6. Once you've placed the image on the screen, you can annotate or add comments using the drawing tools.

Exchange Files ➤ *pp. 158-159*

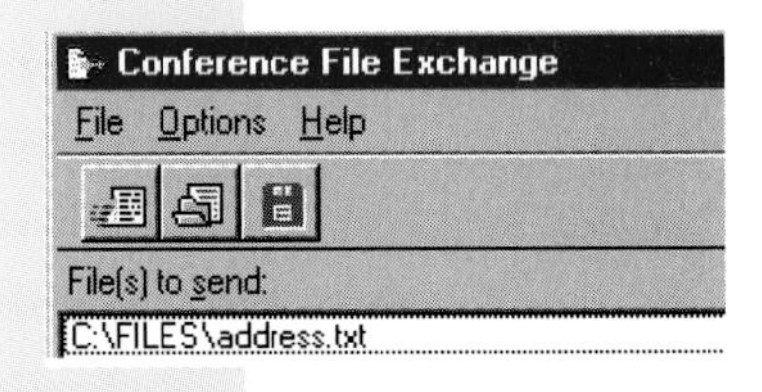

1. Launch Netscape Conference's File Exchange by clicking the File Exchange button on the toolbar or by choosing Communicator | File Exchange.
2. To send a file, you'll need to choose File | Add to send list (or press CTRL-A). Find the file you want to send, click Save, then click the Send icon or choose File | Send.
3. To save any file you've received, highlight the file name, then click the Save button or choose File | Save. Navigate to the folder you want to save the file in and choose Save again.

Chat Using Text ➤ *pp. 159-160*

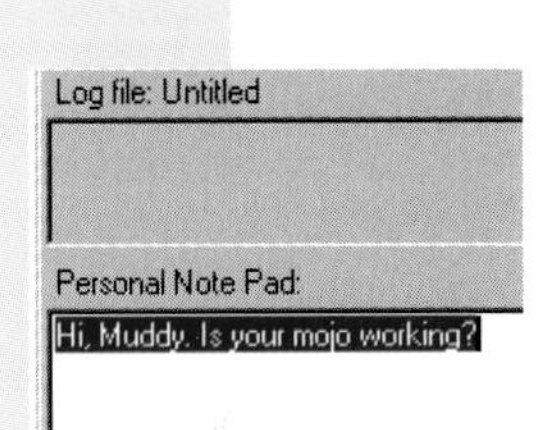

1. Launch the Conference chat by clicking the chat on the toolbar or by choosing Communicator | Chat.
2. Type the stuff you want the other person to see in the bottom box.
3. When you want to send the comment to the other person, press CTRL-ENTER, click the Send icon on the toolbar, or choose File | Post Note Pad.
4. Read the conversation in the top pane.

Send Internet Voice Mail ➤ *pp. 160-161*

1. Choose Communicator | Voice Mail to send voice mail to anyone on the Net who's using Netscape Conference.
2. Type the e-mail address of the person to whom you want to send the voice mail message and choose OK.
3. Click the record button to record your message using your computer's microphone. When you've finished, click the Stop button, then click Send to send the message.

For years, cost-conscious businesspeople have chased an elusive goal: the low-cost, virtual meeting. Of course, you can arrange conference calls and deal with the thorny logistics of who can talk when. Or you can harness expensive video-conferencing facilities—schedule some satellite time, get in a room with a bunch of your colleagues, and talk to a live video image of a bunch of other people somewhere else in the world. And even though these solutions were expensive or inconvenient, they were often better than flying for hours and hours.

But recently, the explosion of the Internet has made this slippery goal achievable. Need to chat for a few minutes with a colleague across the state, nation, or world? No problem. If you've both got sound cards, you can talk back and forth as you would on a speakerphone. Even if you don't, you can collaborate in a real-time, text-based chat.

In this chapter, we'll take a look at Netscape Conference—it's a stealth component of Netscape Communicator 4.0 that's got scads of neat features but is often overlooked in the shadow of Netscape Navigator and Netscape Messenger. We'll also take a brief look at some other Internet-based ways to collaborate, chat, and communicate that don't require Netscape Conference.

Starting a Virtual Meeting

If you've read the book straight through to here, it's going to come as no shock to you to learn that you'll need to set up a thing or two before you use the program.

Is This for Me?

If you're never going to communicate with anyone else on the Web, skip this chapter. But even if you think it's unlikely you'll use this feature, you might want to skim through the chapter anyway—Netscape Conference includes features you may find useful.

Whether you've run the program before or not, you can launch Netscape Conference from the Windows Start Menu by choosing Start | Programs | Netscape Communicator | Netscape Conference.

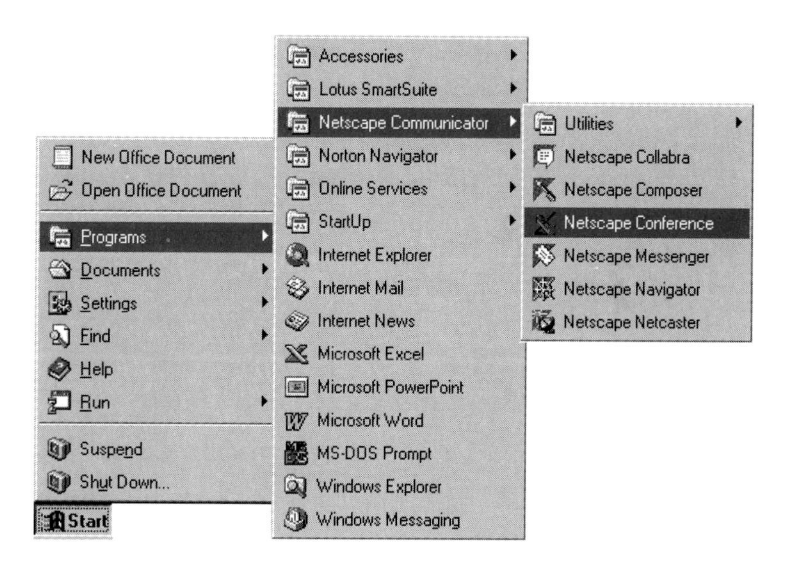

Or, if you're already running another Netscape Communicator 4.0 module, you can choose Communicator | Conference.

Running Netscape Conference for the First Time

The first time you run the program, you'll see this screen:

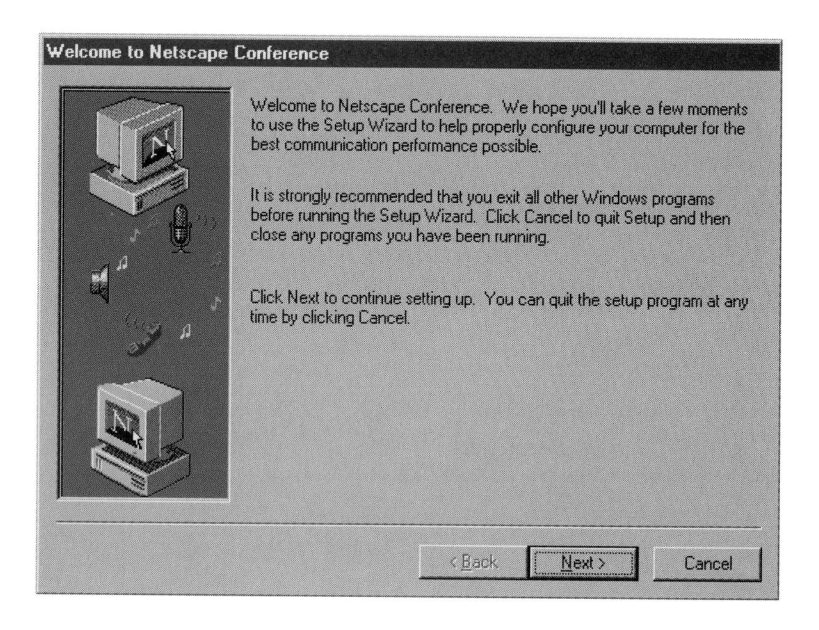

This screen and the one following are pretty easy to follow. Make sure you've closed any other programs and that you're connected to the Net before you continue. As the screens admonish, make sure you have your name and e-mail address handy (although if you forget your name, you've got bigger problems than software to worry about).

Click Next to move to the next screen, then click Next again. You'll come to one of the easiest setup screens you'll ever see. Dive in and fill it out (somehow, those two figures of speech just don't work together, but you get the idea).

1. Type your name.
2. Type your e-mail address.
3. If you have a photo of yourself or another image you want to use to identify yourself, click here, then use the File Open dialog box to find it.

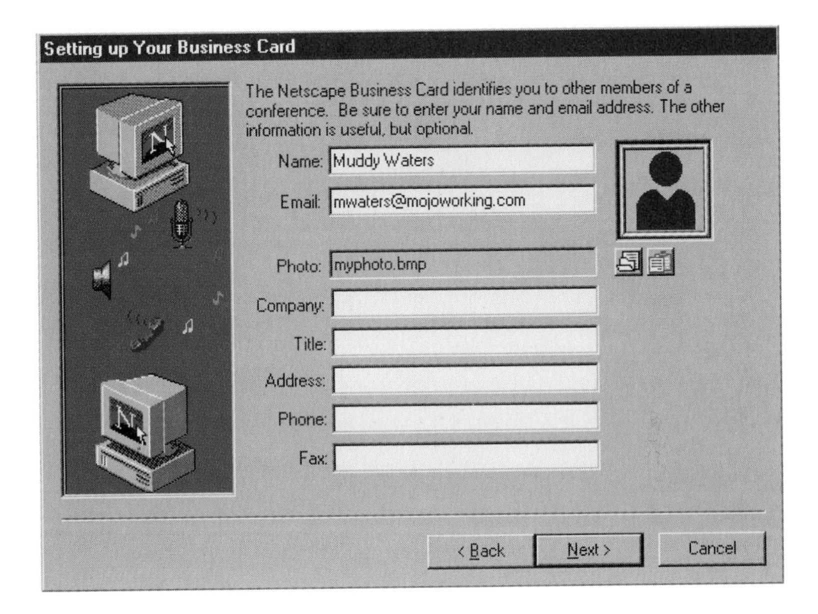

4. Add any other information you would like to include.
5. Click Next.

Once you've typed all your personal information in the dialog box, you'll see a dialog box asking you some confusing stuff about a DLS server. You'll probably want to leave all these settings alone unless someone in your company has told you otherwise.

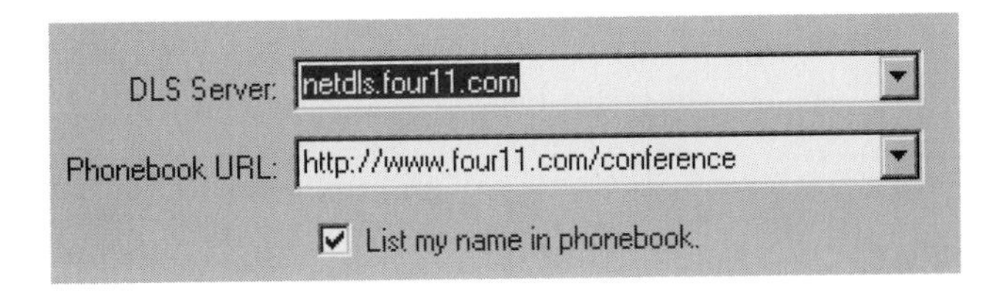

The only thing you might want to change is whether or not you want your name included in the directory—but if most of the people with whom you plan to conference know you have the ability to conference, there's really no advantage to listing yourself in the directory. When you're finished with this information, click Next.

The final bit of information you need to provide is how you're connecting to the Net. Click the button that corresponds to your Internet connection, then choose Next.

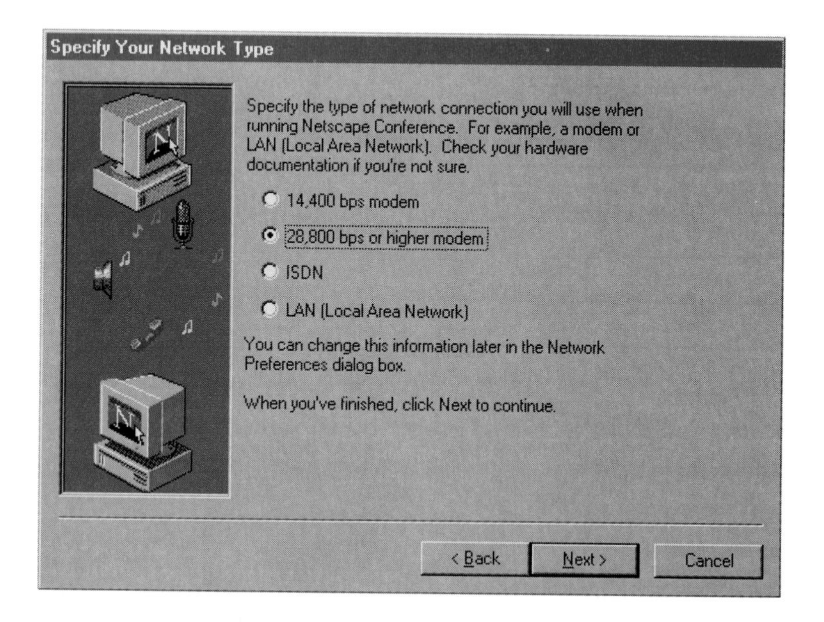

EXPERT ADVICE

If you chose "LAN" as your connection type, you'll see a brief "firewall warning" screen. It'll tell you that the security on your LAN may preclude you from receiving full Netscape Conference abilities, but then again, it might not. Just click Next.

As the final step in the process, you'll need to calibrate your audio settings. If everything's working smoothly, it'll only take a moment. You'll see this dialog box:

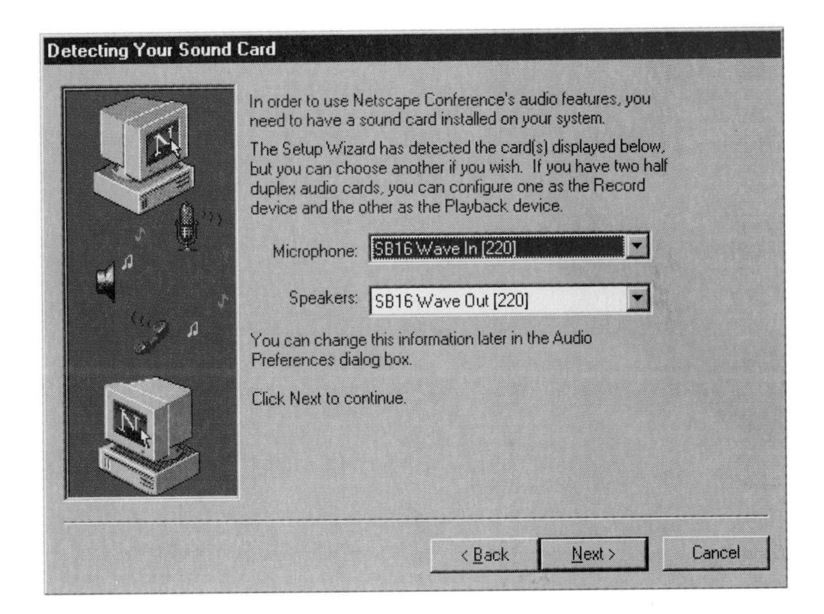

Netscape Conference will automatically determine the audio card you have in your computer. In most cases, the information it shows will be correct—so just click Next. If you're using more than one sound card, you can choose the one you want from the pop-up menus. Once you've clicked Next, you'll see a dialog box telling you that you can test the audio settings. It's a good idea to do so, so choose Next. (If you're just not in the mood for all this, you can choose Skip instead.)

You'll then see this window:

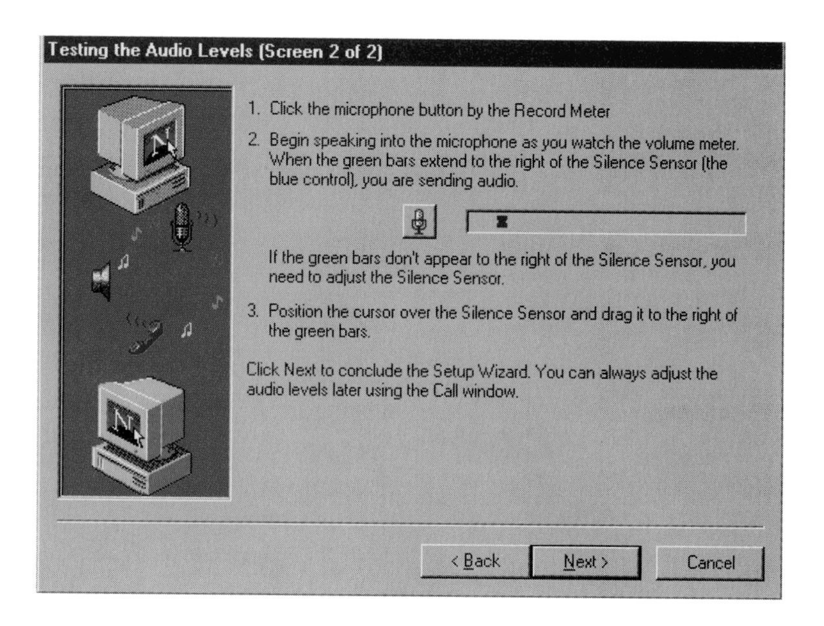

Click the microphone icon and begin to talk in a normal voice. As you talk, green bars should appear to the right of the blue bar. If they appear to the left, drag the blue bar to the left of the green bars. You can play with this all day, but basically, if you see green bars, you're set. Click Next.

At long last, you'll see a screen congratulating you on enduring the lengthy setup process. Click Finish to get on with it.

Running Netscape Conference

Whenever you start Netscape Conference, you'll see the main screen, as shown in Figure 8.1.

From this main window you can access any of Netscape Conference's features.

Making a Call

Before you can do *anything* in Netscape Conference, you'll need to call someone. In that respect, it's like a telephone or fax machine—its value only comes into play when someone else has one.

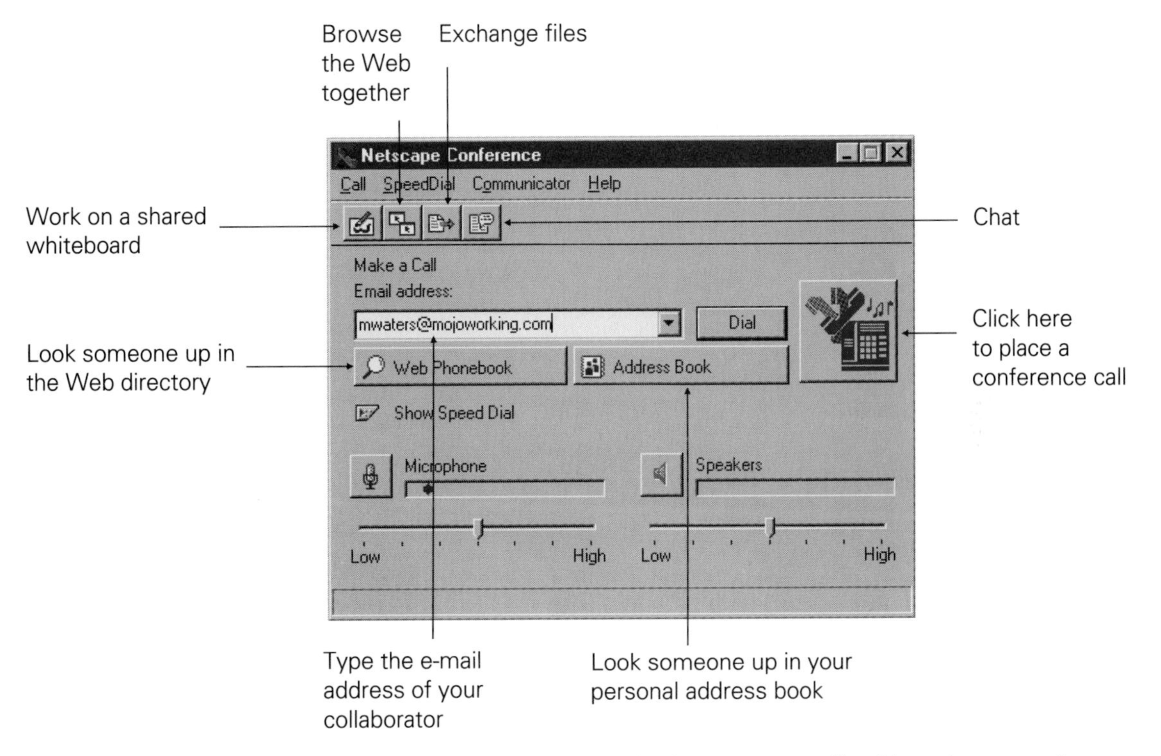

Figure 8.1 The main Netscape Conference screen. Use this as the center of your electronic collaboration

All you'll need in order to call someone is their e-mail address. As long as it's the e-mail address they typed in when they installed Netscape Conference, you'll have no problems.

EXPERT ADVICE

If you call certain people often, you can add them to your Speed Dial list. Click the Speed Dial button in the main window to show a list of speed dial buttons—and then edit your Speed Dial numbers using the Speed Dial pull-down menu.

STEP BY STEP Placing a Netscape Conference Call

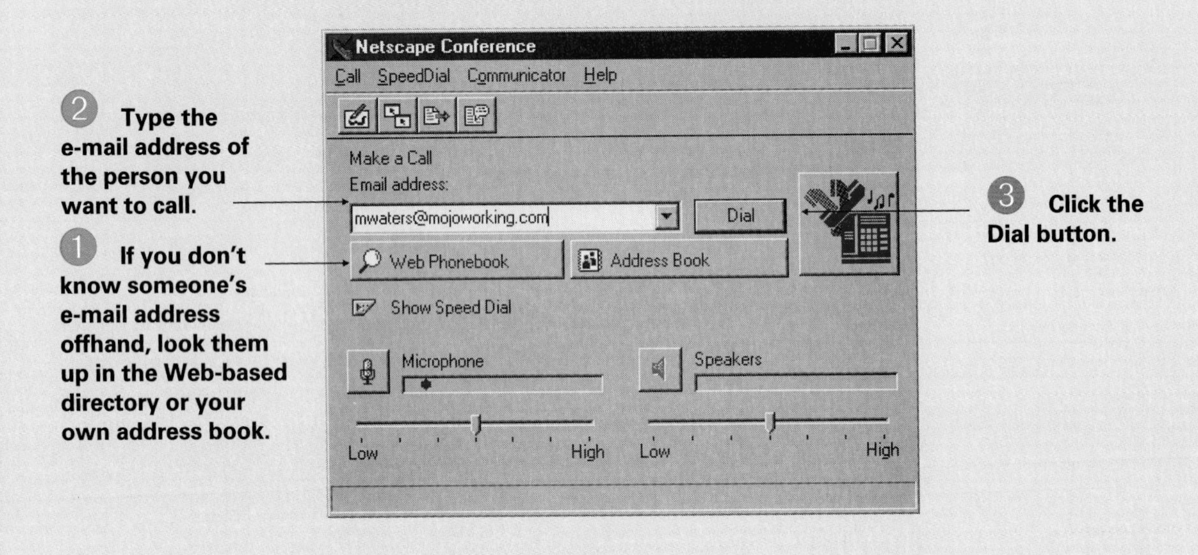

Once you've placed the Conference call, you'll be able to talk with the person you called just as you would on a speakerphone, using the microphone and speakers on your computer.

Collaborating in "Shared Space"

We've all sat in meetings and enjoyed the sharp smell of dry erase markers while someone illustrated points on a whiteboard. You'll have to sniff your own markers, but Netscape Conference provides similar capabilities (Figure 8.2). Launch the shared whiteboard by clicking the whiteboard button on the toolbar or by choosing Communicator | Whiteboard.

You've undoubtedly used some kind of drawing program on your PC before, so all the tools on the left of the screen should look familiar. Remember, you can hold your mouse pointer over any button for a moment and a small label will appear over the button to remind you what it does.

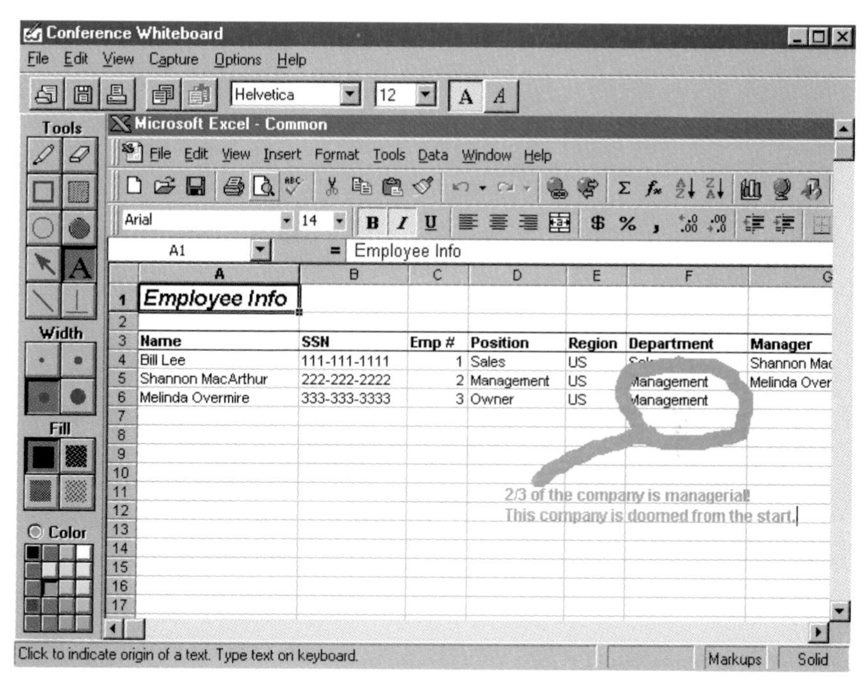

Figure 8.2 Use the shared whiteboard to collaborate and to comment on documents

While drawing together with your collaborator is vaguely amusing, the really useful part of the program comes into play when you use it to annotate something that appears on your screen. To place whatever you want to look at in the whiteboard window, use the Capture menu.

1. Make sure the screen or document you want to capture appears behind your whiteboard window.
2. Choose Capture | Desktop (or Window or Region if you don't want the whole desktop).
3. If you chose Window, click the title bar of the Window you want to capture. If you chose Region, click and drag over the portion of the screen you want to capture.
4. When you're returned to the whiteboard window, position the crosshairs in the window where you want the upper left corner of your screen image to lie.

5. Click once to place the image in the whiteboard screen.

Once you've placed the image on the screen, you can annotate or add comments using the drawing tools (just like the example in Figure 8.2).

Tandem Web Browsing

Sounds like an Olympic event (well, maybe in 2012), doesn't it? And like dancing, you can choose to lead or not. In essence, you'll both be seeing the same thing in a Netscape Navigator window; while one of you browses the Web, you can discuss what you're both seeing.

To launch a shared Web browser, click the shared browser button on the toolbar or choose Communicator | Collaborative Browsing. You'll come to the dialog box shown in Figure 8.3.

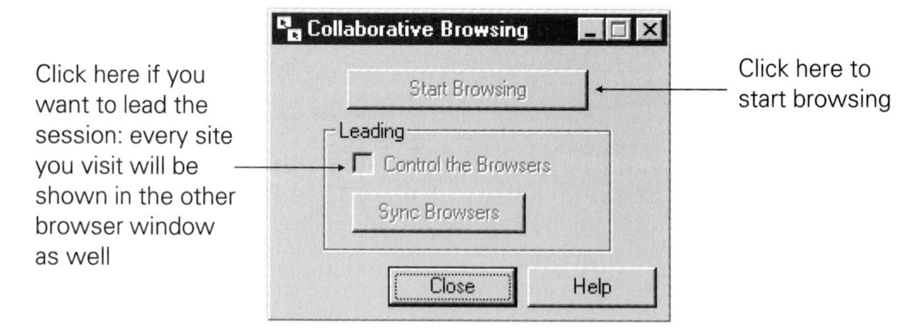

Figure 8.3 Browse the Web with your collaborator

Live E-Mail: Sending and Receiving Files

You'll often find you need to show the people you're working with more than one screen of a document at a time—more than a shared whiteboard can handle. In that case, you'll find it useful to send an entire file, whether it's a graphics image, a word processing document, or an HTML file.

It's easy to send and receive files from your collaborator while you're on a Netscape Conference call. If your collaborator sends you one, you'll see a little pop-up window that tells you so—then simply follow the directions.

To send a file, first launch Netscape Conference's File Exchange by clicking the File Exchange button on the toolbar or by choosing Communicator | File Exchange. You'll then see this window:

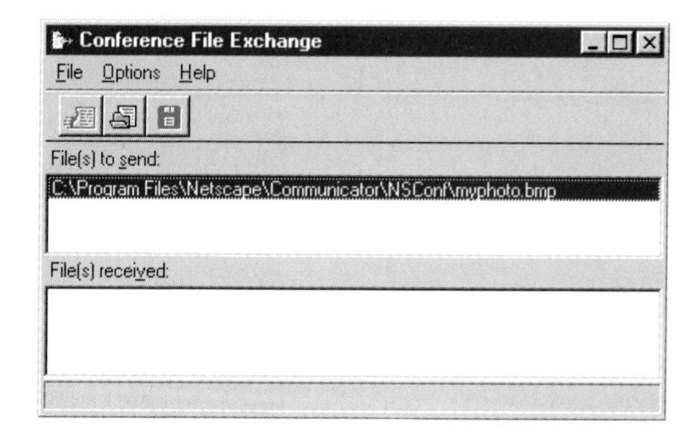

Choose File | Add to send list (or press CTRL-A). Select the file you wish to send and click Open, then click the Send icon or choose File | Send.

In the lower half of the Conference File Exchange window, you'll see a list of files you've received. To save any of those files, highlight the filename, then click the Save button or choose File | Save. Navigate to the folder you want to save the file in and choose Save again.

If Talking Isn't Enough: Text Chat

The fourth major utility in Netscape Conference is a text-based chat. In essence, this enables you to type messages back and forth in real time. We can't see any particularly useful application for this if you're already talking with someone using sound. However, if you want a record of a conversation or you're not using sound to collaborate, a text-based discussion may be just what the doctor ordered.

EXPERT ADVICE

If you're planning a text-based chat, be sure to choose someone who types relatively quickly—or the money you're saving by not using the phone will quickly dissipate compared to the time you're spending waiting for a response.

Launch the Conference chat by clicking the chat button on the toolbar or by choosing Communicator | Chat. You'll see the Chat dialog box shown in Figure 8.4.

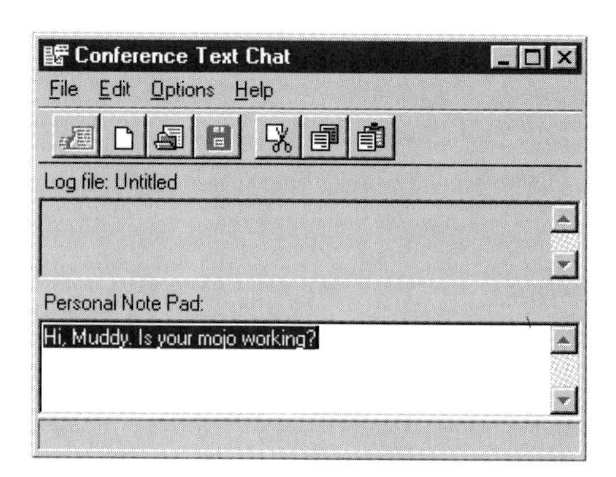

Figure 8.4 You can also engage in a text-based chat

As with the other Netscape Conference modules, this is pretty easy to use. You'll type the stuff you want the other person to see in the bottom box. When you want to send the comment to the other person, press CTRL-ENTER, click the Send icon on the toolbar, or choose File | Post Note Pad.

Your comments and those of the other person will appear in the top pane. Save a transcript of the discussion by choosing File | Save. You can even insert plain text files into your comments—just choose File | Include (or copy and paste text from other applications).

Voice Mail (You Can't Escape It)

Netscape Conference has one last tool that's easy to overlook because it's not listed on the toolbar: Voice Mail. As if having voice mail on your phone system wasn't enough, you can choose Communicator | Voice Mail to send voice mail to anyone on the Net who's using Netscape Conference.

When you launch Conference's voice mail, you'll first be asked for the recipient's e-mail address:

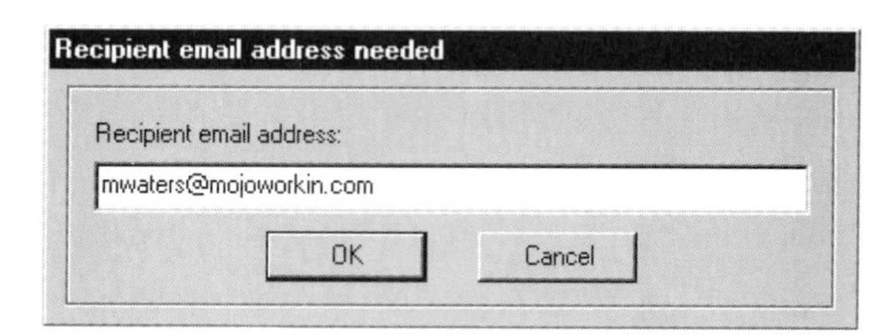

Type the e-mail address of the person to whom you want to send the voice mail message and choose OK. You'll then see this dialog box:

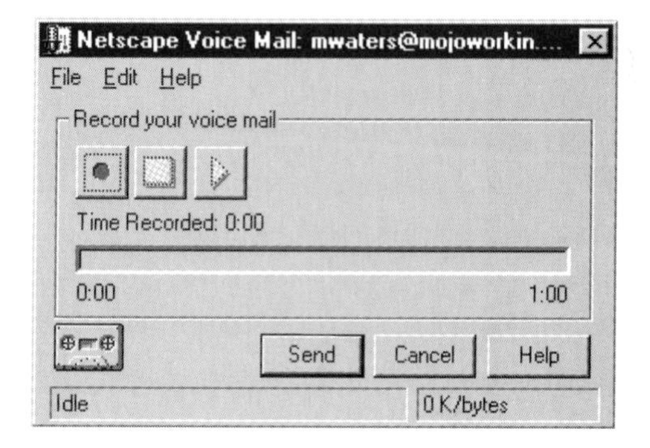

Click the record button to record your message using your computer's microphone. When you've finished, click the Stop button, then click Send to send the message.

The Hired Help: Conference Attendant

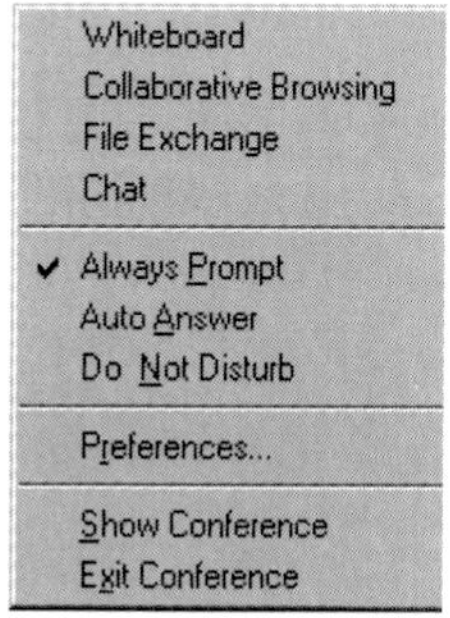

So, how do you know if anyone's trying to start a conference session with you? Easy: if you're currently running Netscape Conference, a handy little utility sits in the background, waiting to see if anyone's calling or using any of the conference tools like the whiteboard, file exchange, or chat.

It's called a Conference attendant, and it rests in your Windows 95 Taskbar, appearing as an icon of two phones at the far side from the Start button. Right-click on the button and you'll get a few more options, like those shown at the left.

You can choose not to be notified if someone's trying to start a conference with you, to automatically have any Conference call answered, or—the default setting—to be asked if you want to answer each time anyone calls.

You can also choose Preferences from this pop-up menu to adjust any of the information you provided when you initially ran Netscape Conference (Figure 8.5).

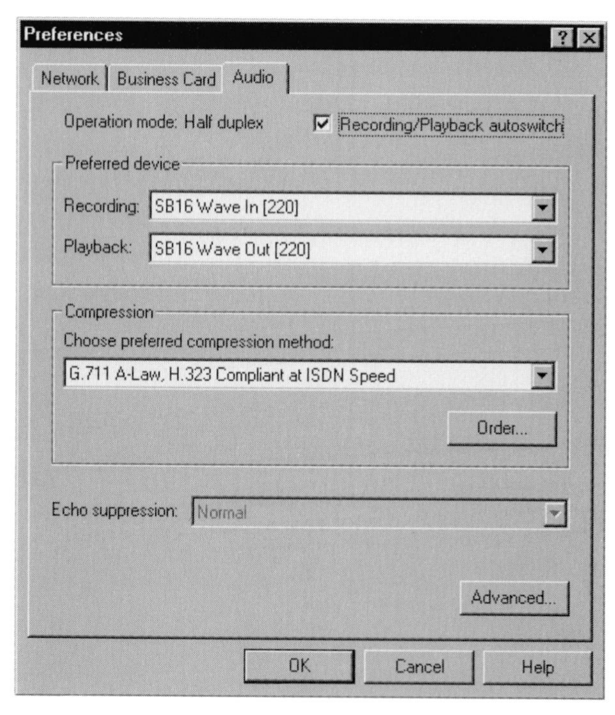

Figure 8.5 Use the Preferences window to adjust your Netscape Conference settings

Other Internet Collaboration

When you're using a new tool like Netscape Conference, it's easy to overlook the fact that many conference and collaboration tools already exist on the Web. For example, text-chat already exists on online services such as America Online and on many Web sites. Some Web sites use chat as an aside to their main content; others, like **www.talkcity.com** and **www.talk.com,** make it the core of their purpose. Still other Web sites include message boards and other conferencing abilities. You may want to explore some of these, but they differ so widely that it's impossible to explore them all here.

After reading this chapter, you should have a good idea of the things you can do using Netscape Conference. Find a friend with Netscape Communicator 4.0 and give it a try!

CHAPTER

9

Creating Web Pages with Netscape Composer

INCLUDES

- Deciding what to publish for which people
- Planning ahead while you can
- Creating an instant home page
- Making a home page with Netscape templates
- Making a home page from a blank document
- Converting and editing existing documents

What's a Web Site For? ➤ *pp. 169-171*

A Web site is the flagship of an organization's Internet presence, the hub of an information center about the organization. A well-designed, effective site is well-planned and well-implemented.

What Is HTML? ➤ *pp. 172-173*

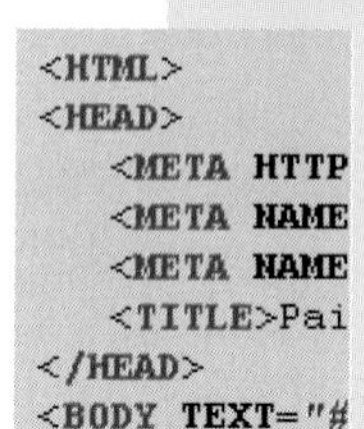

- HTML stands for hypertext markup language.
- HTML is a set of codes that instructs a Web browser both how to display a document and what to do when "hot" links are selected (or clicked).
- HTML describes how a document is structured, not precisely how it should be displayed, so different programs running on different types of computers handle the same HTML codes differently—but they're still faithful to the intentions behind the codes.
- There are several different versions of HTML, most notably HTML 2.0 (the most commonly "understood" version), Netscape extensions to HTML, HTML 3.2, and HTML 4.0 (which is still in the works).

HTML Basics: What's a Tag? ➤ *pp. 174-175*

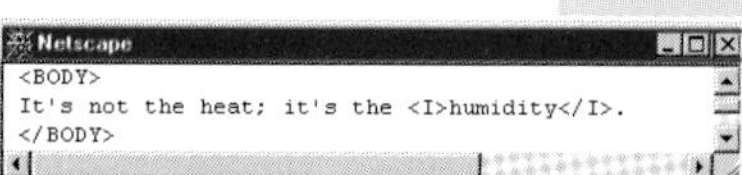

A tag is a set of (usually) paired "codes" contained in angle brackets that tell Netscape Navigator or other browsers how to format text or other page elements.

Launching Netscape Composer ➤ *p. 176*

1. Start Netscape Communicator.
2. Select Communicator | Page Composer, *or*
3. From the Windows 95 Start menu, choose Programs | Netscape Communicator | Netscape Composer.

Make a Home Page with the Netscape Wizard ➤ *pp. 178-184*

1. While connected to the Internet, select File | New | Page From Wizard.
2. Follow the instructions presented by Netscape in the upper-right and left frames.
3. When you're done, click the Build button.
4. Click File | Edit Page.
5. Customize and save your new home page.

NETSCAPE PAGE WIZARD

Use a Netscape Home Page Template ➤ *pp. 184-187*

1. Select File | New | Page From Template.
2. Click Choose File to select a template file of your own, or choose Netscape Templates to choose one of theirs.
3. Click File | Save, name your file, and choose Save.
4. Edit and customize the template.

Choose samples from Netscape's Template Website

Netscape Templates

Start a Blank Web Document ➤ *p. 187*

1. Select File | New | Blank Page.
2. Start typing and formatting your new document.

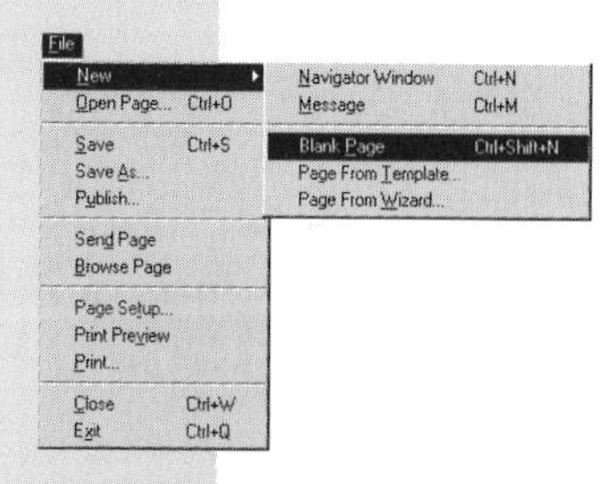

Open an Existing HTML Document in Netscape Composer ➤ *pp. 187-188*

1. In Netscape Composer, select File | Open Page.
2. Browse to the document you want and open it.
3. Save the new copy of the document with a different name.

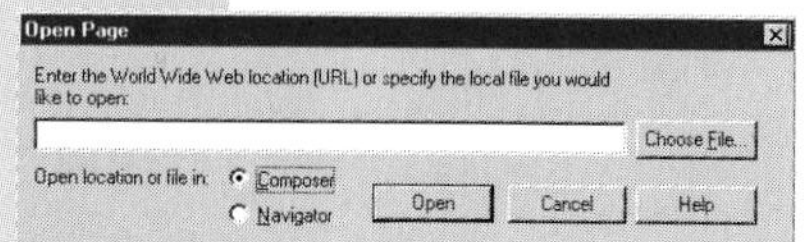

By this point in the book, you've got a pretty clear idea of what Netscape Communicator 4.0 is: it's a collection of tools that help you work on the Web, yadda yadda yadda. Well, if you're creating Web pages, you'll use Netscape Composer. This chapter—and the three following it—help you learn the basics of Netscape Composer. Don't forget: many of the formatting abilities you'll learn about in Netscape Composer can be applied when you're creating e-mail messages, too, as long as the person you're sending them to can receive HTML-formatted e-mail.

EXPERT ADVICE

Don't be intimidated by the prospect of creating Web pages. If you've ever used a word processor, you'll already be familiar with many of the tools in Netscape Composer. It'll be old hat to you before you know it.

Is This for Me?

If you know for certain that you're never, ever going to create a Web page, you can skip the rest of the book. But you'd be wise to at least *skim* the remaining chapters, because it's a good idea to know how Web pages work, even if you never plan to build one. (This same logic explains Jeff's Dad's constant attempts to make him understand how his VW Bug ran while he was a teenager.)

Planning Ahead: Site and Pages

You're reading this book because you don't have time to read for hours in order to create a Web page. In this and the following chapters, we'll sit down together to learn the basics of creating Web pages. With Netscape Communicator 4.0, you've already got a full complement of tools to help you create attractive, compelling Web sites in a flash.

Before You Start

Whether your site will be big or small, bare-bones or fancy-pants, you'll still have to answer some key questions. To keep your efforts focused and on track, keep your answers to the following questions in mind as you create your Web site and pages:

- How will you get people to come and visit?
- How will you convince them to stay?
- Why will they want to come back?
- What's your purpose—why a Web site?
- What information will you put out?
- How will you imbue your site with your company's personality?
- How often will you change or update the site?
- How interactive will your site be?
- How much time and effort can you devote to the project?

EXPERT ADVICE

The terms "home page" and "Web site" are beginning to have different connotations. "Home page" now suggests a personal page or solely the main "hub" page of a larger site. The term "Web site" suggests a fully fleshed out complex of Web pages. If you're planning to create a credible business presence, think in terms of a full-fledged Web site.

Your most important decision? To determine your Web site's purpose. A business, for example, might use their site to support marketing and public relations or as part of advertising efforts. Other companies may use the Web to provide support to customers or even to sell products.

Thinking It Through: What's Your Metaphor?

Whatever purpose you identify, you'll also want to choose a "metaphor," or a way to approach your goals. The Web is so new that there's no set way to arrange information, but you need to approach it in a way that's familiar in order to make it coherent. Many businesses choose a relatively traditional publishing approach, whether or not they're actually in the publishing business. (For a look at how one newspaper publisher approaches their Web site, see Figure 9.1.)

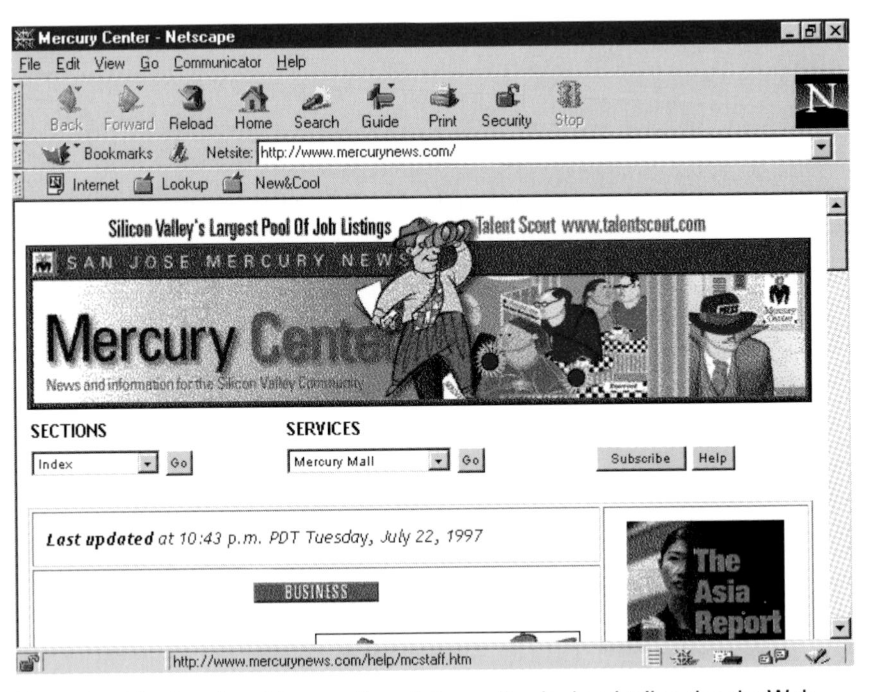

Figure 9.1 The San Jose Mercury News' site, well-suited and tailored to the Web

Something about computers makes people feel like they need to use acronyms and goofy jargon. Don't be put off. Instead, learn a few basics, like those mentioned throughout the book. You'll then know plenty to get by—and when you come across a new term, just figure it out in context or look it up using one of the resources listed in Appendix A.

Of course, a Web site isn't a newspaper or magazine—you can update it every day, and your visitors don't have to wait for it to appear in the mailbox. A Web site isn't "physical," either, but you might choose to create one that resembles your company's physical presence, as did the Gap (Figure 9.2).

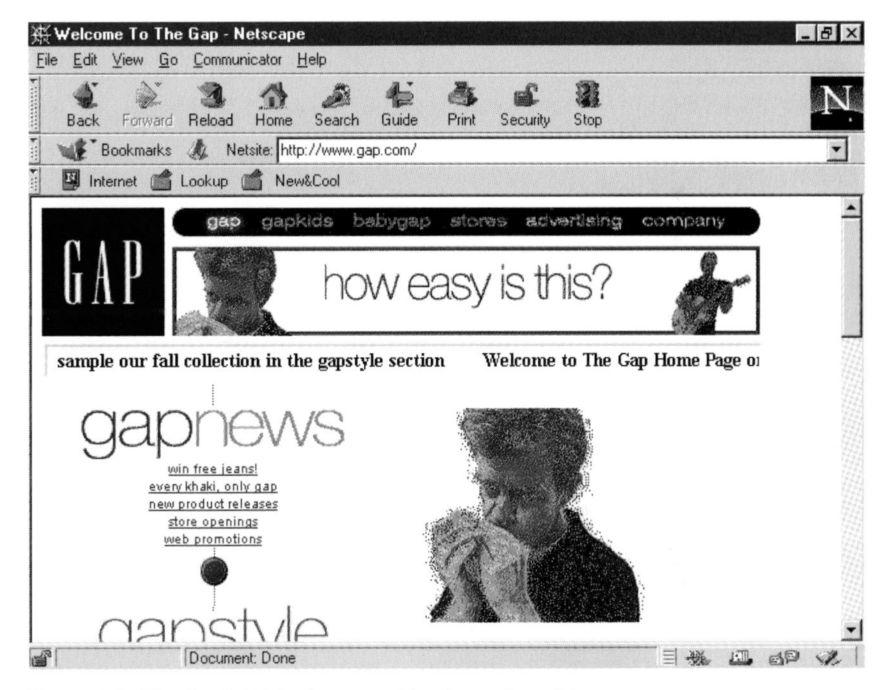

Figure 9.2 The Gap's Web site resembles its real-world stores

Getting Started with Web Page Design

Enough navel-gazing. You're anxious to move past all this introspection, so let's get started. In the rest of this chapter, you'll learn some of the basics of HTML (the building blocks of Web pages). You'll learn how to create a Web page from scratch, plus you'll produce instant pages using Netscape Communicator 4.0's templates. In later chapters, you'll master the fundamentals of Web pages, including formatting text, adding graphics, adding hyperlinks, and more.

What Is HTML, Anyway?

With a Web-publishing program like Netscape Composer, you don't have to "dirty your hands" typing in HTML codes, but by being aware of HTML and how it works, you'll get a clearer idea of what Netscape is really doing when you edit and format your documents.

Is This for Me?

You don't *need* to know about HTML codes, but, again, it'll help you understand what Netscape Composer's doing behind the scenes. Trust us: we'll make it brief.

HTML stands for *hypertext markup language.* It provides a set of codes that instructs Web browsers both how to display a text document (the document's structure) and what to do with page elements (like graphics and hyperlinks). When Netscape Navigator or another Web browser "opens" a Web page, it translates these codes—called "tags"—and displays the document onscreen as the tags dictate. (For a look at a simple Web page and the codes used to create it, see Figures 9.3 and 9.4).

SHORTCUT

Want a quick way to learn how a particular Web page is created? Load the page in Netscape Navigator 4.0, then choose View | Page Source, or press CTRL-U. A new window will appear, showing you the HTML codes and tags behind the page layout.

Understanding HTML codes gets a little easier after a while. Nevertheless, you shouldn't feel like you're missing out on anything all that great if you avoid dealing with HTML directly.

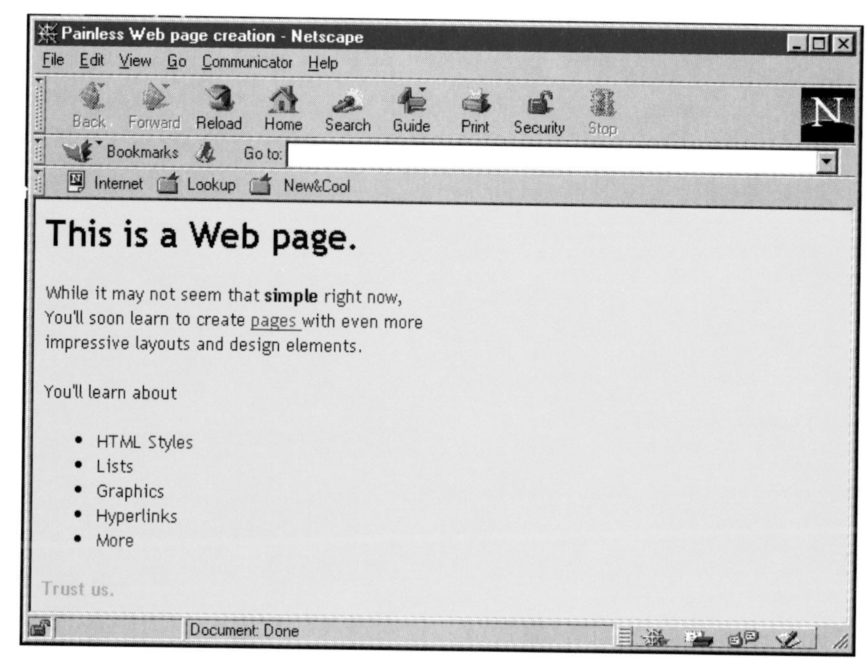

Figure 9.3 A simple Web page

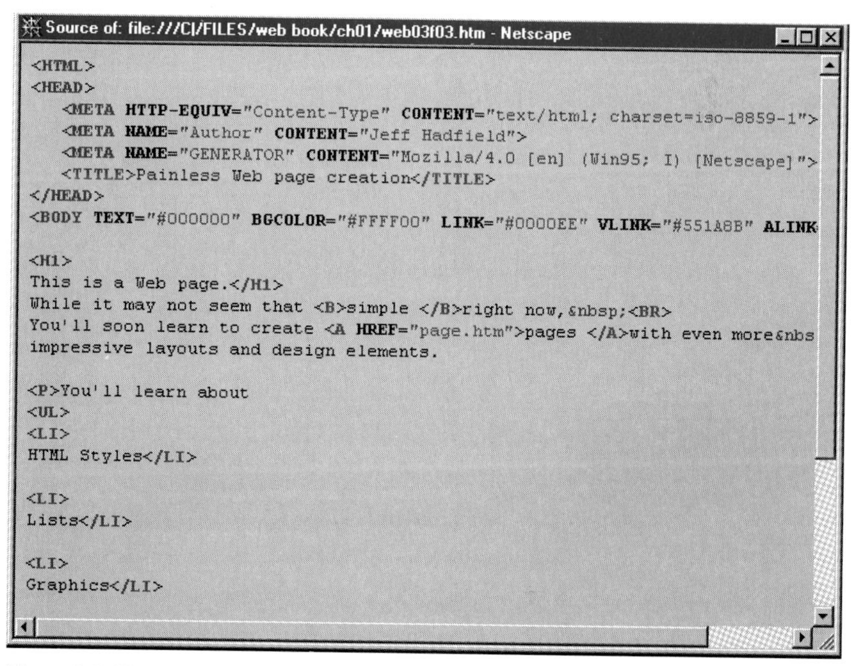

Figure 9.4 The HTML code, including tags, used to create the page shown in Figure 9.3

A Little More about Tags

Most of the codes used in HTML are *tags*. Tags appear between angle brackets: <>. Tags can be created with upper- or lowercase letters. For example, both <p> and <P> have the same effect: they indicate the start of a new paragraph.

Most Tags Mark Text by Surrounding It

Most tags occur in pairs and surround the text they affect. For example, here's a portion of an HTML source document:

```
To make a word appear <B>darker</B> and more prominent,
you can surround it with the HTML bold tag.
```

And here's what the document looks like when it is viewed with a Web browser:

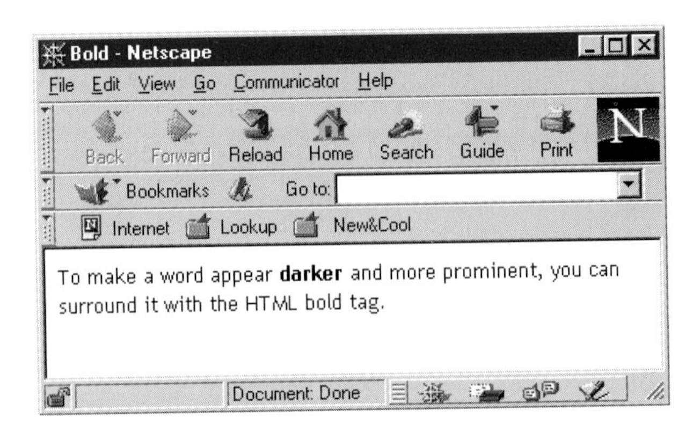

The <B> tag before the word *darker* applies boldface to the word. The </B> tag indicates the end of the boldfacing.

Some tags, most notably the line break tag, don't surround text. But for the most part, HTML tags occur in pairs.

CAUTION

Although you won't type these tags directly into Netscape Composer, it's good to know how they work so you can better troubleshoot your HTML documents.

More Than One Kind of HTML

As if all that wasn't confusing enough, you'll also be hit with different versions of HTML. The most widely accepted standard is HTML 2.0, although the more current HTML 3.2 is supported by nearly every major browser, including Netscape Navigator 4.0. Of course, there's a new proposal out for HTML 4.0, so, just like the rest of the Web, HTML's capabilities will keep increasing.

The scary part: Netscape and Microsoft don't always agree on exactly how to implement HTML in their browsers, so you may find some incompatibilities here and there between the two. But the fact is, unless you're performing some highfalutin' hijinks or consciously *trying* to use a Netscape-only feature, you probably won't run into any problems creating basic pages.

EXPERT ADVICE

On the Web, imitation is the sincerest form of flattery. While you don't want to steal exact coding, you can pick up page-formatting tips by viewing the source of Web pages with layouts you like.

Now it's time to build your Web site!

Launching Netscape Composer

Like every other part of Netscape Communicator, launching the Composer module is as easy as pie.

Is This for Me?

This section—and the rest of the chapter—show you what you need to know about running the basic features of Netscape Composer. No matter what you're planning to use Netscape Composer for, you should read the rest of the chapter.

Here's how:

1. Start Netscape Communicator.
2. Select Communicator | Page Composer, *or*
3. From the Windows 95 Start menu, choose Programs | Netscape Communicator | Netscape Composer.

Or, if you're feeling zippy, you can just click the Composer icon on the far right (or bottom) side of the Component Bar.

Making a Web Document (for Busy People)

Let's get down to business. While it used to take armloads of software to create Web pages, you've got all you need to get started in Netscape Composer. We'll ignore, for now, the planning, staffing, and organizing aspects and give you our magic, four-step process.

Four Simple Steps

Using Netscape Composer, you can create a Web page in four basic steps:

1. Create or open the document.
2. Edit and format the document.

3. Save the document.
4. Publish the document.

Create or Open the Document in Netscape Composer

You can create a Netscape Composer document from scratch or open an existing one, whether it's in HTML or another format.

Creating Documents from a Mix or Scratch Like using packaged mix for your brownies, Netscape provides two shortcuts to get you started with new documents. The full-service option is Netscape's Page Wizard, explained step-by-step in the next section, "But Wait! I Just Want a Simple Home Page." Want more varied options? Try Netscape's Web page templates.

If you're feeling like measuring flour, sugar, etc. on your own, you can start from scratch with a blank document. Type or insert text, format your document, add links and pictures, and you're done.

Edit the Document

After opening or starting a document, you'll need to fill in the text you want (or, if you're using templates, replace the placeholder text with real stuff). You can paste existing text from documents created in other applications, if you like. The next few chapters will fill in the details on typing, editing, formatting, hyperlinking, and more advanced design options.

Save the Document

When you're done editing the document, as with any other type of application, you'll want to save it. This is not the same thing as "publishing" the document. *Saving* a document means updating the copy stored on your computer; *publishing* it means posting a copy of it to the Web server, where it's available to the public.

Publish the Document

As mentioned in the previous section, publishing a document means transferring a copy of it to the Web server, where it becomes part of a public site. Instead of using a separate program (as we used to in the "old" days), Netscape

Composer's one-button publishing posts a document (or a full set of documents) directly to a site.

CAUTION

It's a good idea to schedule a testing period after publishing the site but before publicizing it, so that a small group of colleagues or trusted clients can visit the site and look for errors or problems. When the testing period is concluded, you'll want to announce the page or site in the various search utilities and in other public forums on the Internet, when appropriate.

The Big Picture

We want to be careful not to blur the distinction between the fairly simple process of creating a single Web page and the more complicated endeavor of assembling and posting an entire Web site. The steps involved in putting together a site are similar to those spelled out in the previous sections but involve some overarching considerations. For an entire site, you should think in terms of these steps:

1. Plan the site (and think about who will maintain it and how updates will be made).
2. Assemble the documents.
3. Edit and save all the documents.
4. Publish the documents.
5. Promote the site.

But Wait! I Just Want a Simple Home Page

Okay, okay: you don't have to plan a million-dollar Web site before you can produce a simple home page. The easiest way to crank one out is with Netscape's Page Wizard. The Page Wizard is actually stored at the Netscape Web site, not on your computer, so Netscape can change it or improve it whenever they want. When you start a document based on the Page Wizard, Netscape Communicator 4.0 connects you to the Page Wizard...um...*page* and guides you through creating a page.

When you've finished filling out the Page Wizard form, the Wizard makes your page for you—then you just save it and modify it however you like.

Here's how to use the Page Wizard:

1. To begin, select File | New | Page From Wizard.

   This connects you to the Netscape Page Wizard page, which is divided into three frames, two of which are blank initially (see Figure 9.5).
2. Read the brief introduction in the nonblank working frame, scrolling to see the entire text.
3. When you're ready, click the Start button.

Instructions will appear in the upper-left frame, while the upper-right frame turns into a Preview area for the page you'll create (See Figure 9.6).

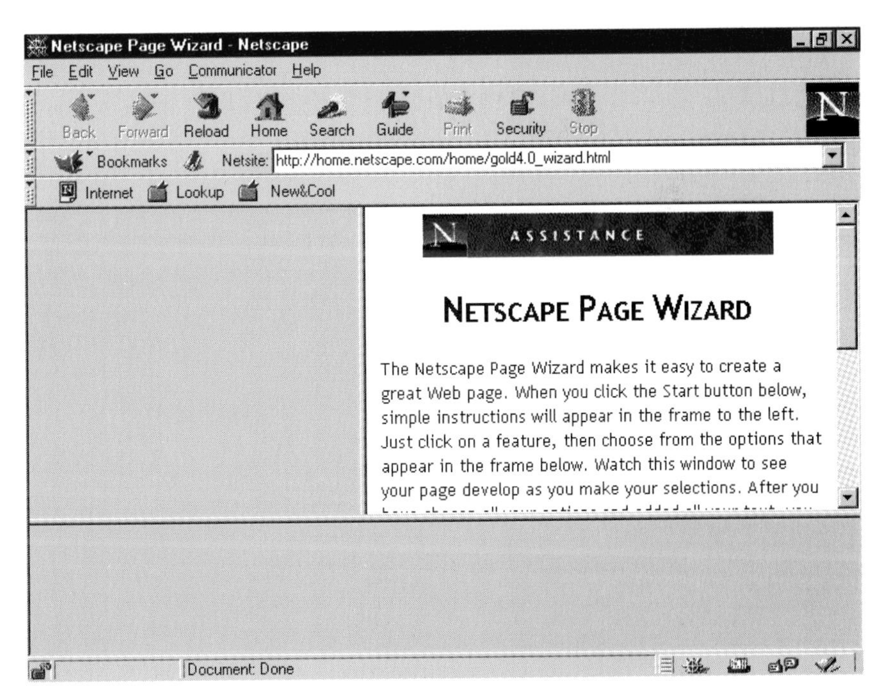

Figure 9.5 Netscape Communicator 4.0's Web Page Wizard

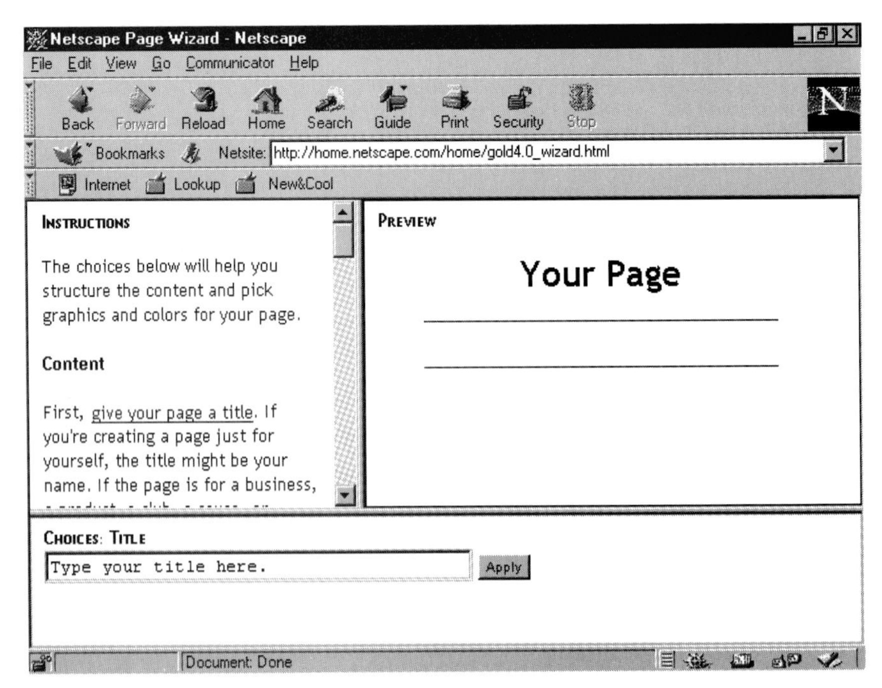

Figure 9.6 Follow the instructions in the left frame—and view the results in the upper right frame

Adding Content to Your Page

First, the Wizard prompts you to supply the content for your page.

1. Click "give your page a title."

A box appears in the bottom frame.

2. Type a title in the box (replacing the dummy title text).
3. Click the Apply button.

   Your title appears in the Preview area.
4. Click "type an introduction" in the upper-left frame.

A slightly larger box appears in the bottom frame.

5. Type something about yourself or your organization in the box.
6. Click Apply.

   The introduction you typed appears on the Preview of your page.
7. Click "add some hotlinks to other Web pages."

   You'll see two boxes in the bottom frame:

CHOICES: HOT LINKS Delete Hot Links

Name: Type the name of your hot link here.

URL: http://your.url/goes/here.html Apply

8. Type the name of one of your favorite Web pages in the Name box.
9. Type that page's URL (Web address) in the URL box.
10. Click Apply.
11. Repeat as often as you wish.
12. Click "type a paragraph of text to serve as a conclusion" in the upper-left frame.

    A box appears in the bottom frame.
13. Type a little more about yourself or your organization in the box.
14. Click Apply.

The conclusion appears on the Preview of your page in the upper right frame.

If you want people to be able to reach you by e-mail through your page, follow the next few steps:

1. Click "add an email link."
2. Type your e-mail address in the box that appears.
3. Click Apply.

Now the content of your document is complete, and you can spruce up its appearance.

Designing and Formatting Your Page

Page Wizard helps you choose color combinations, background patterns, and fancy line and bullet styles for your page.

1. Scroll down to the Looks section of the Instructions in the upper-left frame and read the brief introduction.
2. To choose one of the Wizard's predesigned color combinations, click "a preset color combination."
3. Choose one of the color combinations that appears in the bottom frame:

CHOICES: COLOR COMBINATION

text links visited | text links visited | text links visited | text links visited | text links visited | text links visited | text links visited | text links visited | text links visited | text links visited | text links visited | text links visited | text links visited | text links visited | text links visited

Your page appears in the Preview with the colors you selected. If none of those combinations appeals to you, you can select each element separately by doing the following (skip the next few steps if you like the preset colors just fine):

1. Click "background color."
2. Choose a background color.

Or, if you prefer something fancier:

1. Click "background pattern."
2. Choose a background pattern from the palette:

CHOICES: BACKGROUND PATTERN

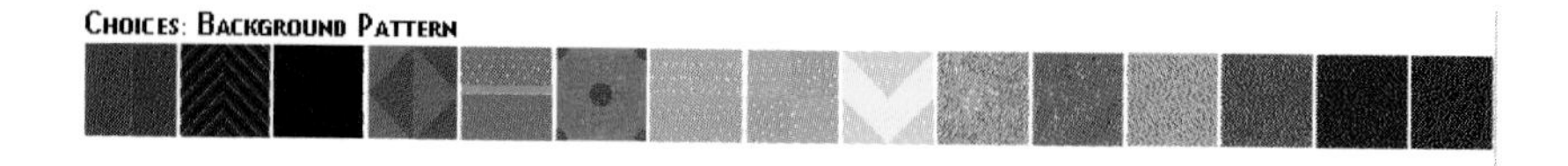

You can also customize the text colors:

1. Click "text color."
2. Choose a text color.
3. Click "link color."
4. Choose a link color.
5. Click "visited link color."
6. Choose a color for links that have been clicked.

Finally, you can select a bullet and horizontal rule style to add a little (very little) panache to your page.

1. Click "choose a bullet style."
2. Select one of the graphical bullet elements offered (notice that a few of them are animated):

CHOICES: BULLET STYLE

3. Click "choose a horizontal rule style."
4. Select one of the graphical lines (you can scroll or increase the size of the frame to see all of the rules).

Finish the Page

If you change your mind about any of your decisions, just click again on the choices in the instructions frame and make different selections until you're satisfied. When you are, you can finish the page:

Build

1. Scroll down the upper-left frame and read the rest of the instructions.
2. Click the Build button.

Your page is assembled (with a free plug for Netscape at the bottom), ready for you to save to your own computer.

From this point on, the procedure is the same as for saving any document from the Web to your local computer for editing with Netscape Composer. Here's what you do:

1. To save the page to your computer, choose File | Edit Page.

   This will open the file in Netscape Composer.
2. Click the Save button.

   The Save As dialog box will appear:

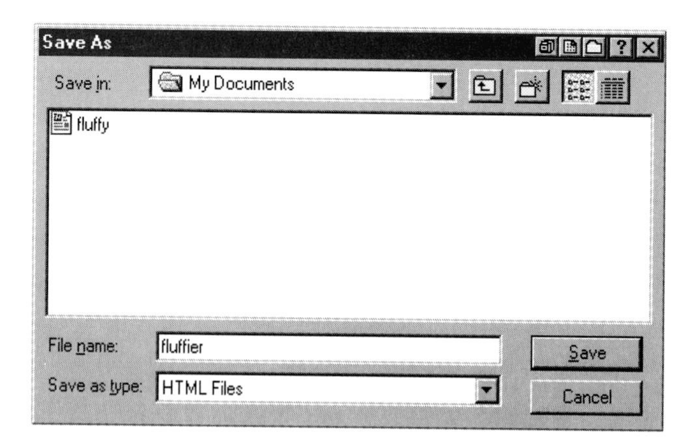

3. Select a folder in which to save this document (and maybe your entire site).
4. Type a file name for the document.
5. Click Save.

Starting a Web Document with a Template

Netscape also provides some document templates for specific purposes. These contain recommended layout and structure as well as boilerplate text you can replace with your own specifics. This is a great way to get your documents started—you can worry about editing them into shape later.

Because of Netscape's simple one-button Edit command, you can also take, as a model, any Web document out there that you like, copy it to your hard disk, and work from it as a template. Just be sure you're not infringing on anyone's copyrights! (It's best to communicate with anyone whose work you're drawing on heavily.)

Let's cover the bona fide templates first.

Starting a Web Document with a Netscape Template

To start a Web document using a Netscape template, first select File | New | Page from Template | Netscape Templates.

This connects you to the Netscape Web Page Templates page. Read the couple of screenfuls of introductory material at the top of the page. Then read the overview of steps for using a template.

Choose a Template

The options include Netscape's Page Wizard (covered in the previous section), along with a number of categories, including

- Personal/Family
- Company/Small Business
- Department
- Product/Service
- Special Interest Group
- Interesting and Fun

Be sure to read each item separately, since some of them don't fit too well in their categories. (For example, we'd put Home Sale Announcement under Personal/Family instead of Company/Small Business.)

To use one of the templates, just do the following:

1. Select and click a template name. Navigator will take you to the template.
2. Choose File | Edit Page.
3. Click the Save button in Netscape Composer. The Save As dialog box will appear.
4. Select a folder in which to save this document (and maybe your entire site).
5. Type a file name for the document.
6. Click Save.

Edit and Clean Up the Document

Edit and customize the template to suit your needs. For example, start by removing the warning (but read it first!). Click and drag to select the horizontal rule at the top of the page and the text beneath it as you would select text and graphics in a word processor.

Then press DELETE. Voila! The line and text disappear. If you change your mind, select Edit | Undo or press CTRL-Z to undo your most recent action. For more on typing, editing, and formatting text, see the next chapter. Rewrite and hack that document into shape. Paste in text from other documents if you need to.

Leave the dummy template art in the document at first as placeholders, so you can replace them with your own art later without having to set things like the alignment options yourself.

Using Any Web Document as a Template

We used to learn HTML by using the browser's View Source command to see the code underlying the pages we liked. Now "borrowing" other people's design ideas is even easier, since all you have to do is download a copy of what you're viewing, save it, and start modifying it.

Say you wanted to base your home page on the Whirled-Wide Med home page (at **http://syx.com/x/busy/wwm.html**). You could point Netscape Navigator at that address and get the page on your screen (see Figure 9.7).

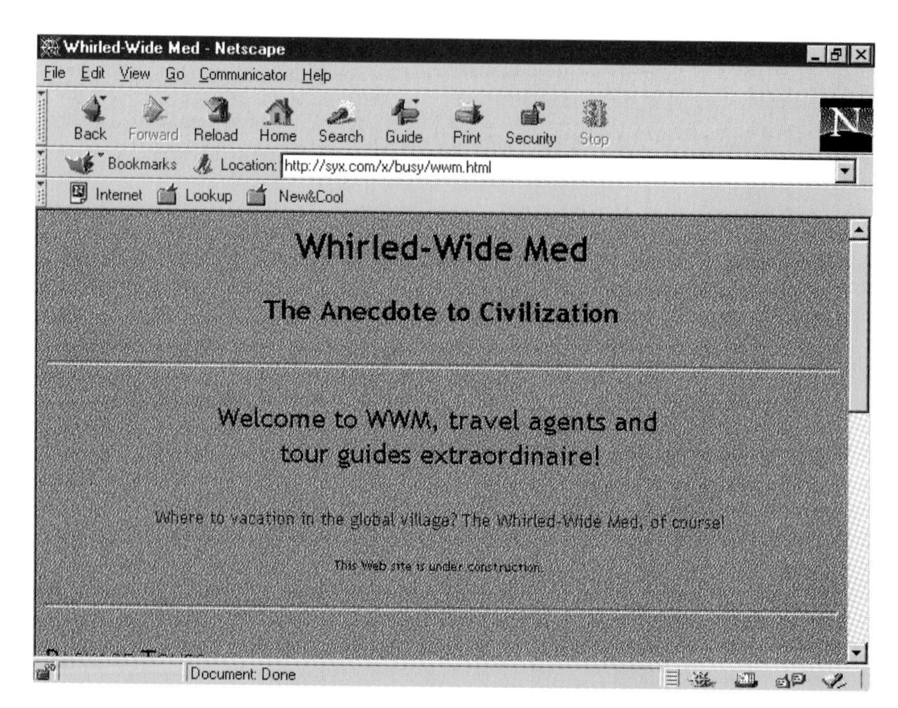

Figure 9.7 You can use Whirled-Wide Med's home page as a model for your own

Now, to start working on your own copy of this page, just do the following:

1. Choose File | Edit Page. The page will be opened in Netscape Composer.
2. Click Save. The Save dialog box will appear.
3. Select a folder to save this document (and maybe your entire site) in.
4. Type a file name for the document.
5. Click Save.

CAUTION

Don't use art, text, sounds, or anything without permission. Not only is it impolite, but it's often illegal and in violation of copyright!

Does this sound familiar? It's the exact same process used to download "real" templates from the Netscape site. From this point on, follow the same advice we gave you in the previous section!

Starting with a Blank Slate

It *is* possible to face up to that proverbial blank white page and just write your own Web document without cribbing other people's ideas or using them as a boilerplate. To do so, select File | New | Blank Page.

Netscape Composer will create a new document named Untitled and let you go at it. Jump ahead to the next chapter to start typing, editing, and formatting your document.

Opening an Existing Local Document

If you have a document you've already started working on (even one you've created in another Web editor or a plain-text editor), you can open it in Netscape Composer and handle it the way you do your new pages.

1. To open an existing Web document, select File | Open Page and select the Composer radio button:

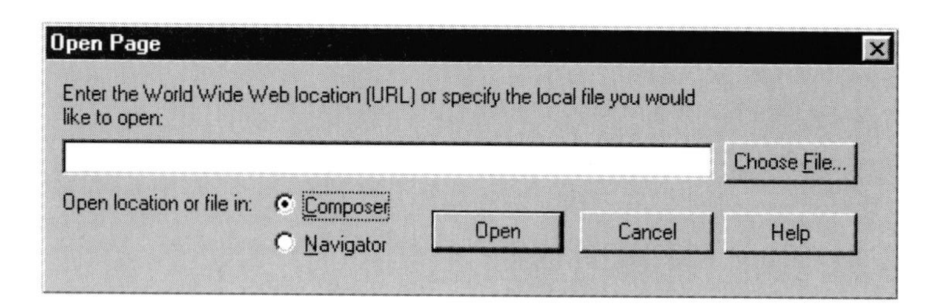

2. Browse to and open the document you want. It will appear in the editor window.

Be sure to save any new copy of a document (after you've opened it in Netscape Composer) with a different name in case something doesn't come out the way you want. That way you've still got your original version of the document.

Converting Non-Web Documents

Often when you start building a Web site, you already have existing documents (or images) that you plan to adapt to the Web. If the documents are already designed and formatted (as opposed to memos and draft text), then you'll want to preserve as much of the formatting as possible so you don't have to do it again. Fortunately, the formatting HTML tags correspond to a fairly standard set of formatting features that most programs can handle. (In fact, word processing and desktop publishing programs usually have more text formatting features than the Web does.)

This means that you can often convert the existing documents into at least partially formatted Web documents. You'll still need to insert the hyperlinks yourself. Also, no conversion tools or programs work perfectly, so you may end up having to reproduce some of the formatting in Netscape Composer.

For that matter, not all document formats can be easily converted. You may need to convert documents into some more universally readable interim format, such as Word for Windows 2.0 or RTF (rich text format), and then convert that into HTML. As a final resort, you can always convert written documents to "plain text" format (usually with the extension .txt). You will lose all of the formatting, but at least you won't have to retype the text. (Most programs that handle documents have a "save as text" option, often on a list in the Save As dialog box.)

Finding Converters

Depending on the file formats you're starting with and the type of computers you use, you'll need to explore different conversion options. For example, regardless of your platform (Mac, UNIX, Windows), both Word for Windows 2.0 (or RTF) and WordPerfect 5.1 file formats are readable by just about any word processing or page layout application.

Of course, the latest versions of both Microsoft Word and Corel WordPerfect (for Windows 95) include integrated, powerful HTML layout abilities. If you're using a previous version of either program, you may be able to download add-in modules from either Microsoft (**www.microsoft.com/office**) or Corel (**www.corel.com/wordperfect**).

EXPERT ADVICE

To prepare a plain text file, first copy it to a new document with a file extension of .html (or .htm). Then open the document in Netscape Composer and start editing it.

If you can get your documents into RTF (rich text format), there are two ways you can convert them to HTML. One is to open them in a word processor that can save in HTML format (such as Microsoft Word or Corel WordPerfect). The other is to convert them directly with an RTF-to-HTML conversion utility. There are different programs for different types of computers (often with similar sounding names—for example, there's RTFtoHTML for the Macintosh, rtftohtml for UNIX, and rtf2html for Windows). See Appendix A for sites where you can download these and similar converters.

EXPERT ADVICE

*If you're planning to convert a large number of documents in a "batch" or bring together a lot of documents in a mess of different formats, you may want to try an industrial-strength, commercial document conversion package. Jeff's favorite: Conversions Plus from DataViz (**www.dataviz.com**).*

Once you've started a new page, you have to fill it with information, format it, and link it to other pages. The next few chapters will help you do just that.

CHAPTER

10

Formatting and Sprucing Up Your Web pages

INCLUDES

- Titling a document
- Typing text in your Web page
- Saving documents
- Editing your document
- Formatting for structure, design, and clarity
- Previewing your Web page
- Inserting hyperlinks

FAST FORWARD

Title a Web Document ➤ pp. 197-198

When Bad Things Happen to Craz : file:///Untitled - Netscape Composer

1. Select Format | Page Colors and Properties and click the General tab.
2. Type a title in the Title box.
3. Click OK.

Type Text in a Web Document ➤ pp. 198-199

Welcome to my page!

- Just type.
- Let Netscape handle word-wrapping.
- Press ENTER to start a new paragraph.

Save a Web Document ➤ pp. 199-200

Press the Save button.

Format Your Document ➤ pp. 203-210

Make a selection and then format it:

- Select heading levels from the Paragraph style drop-down list.
- Click the Bold, Italic, or Underline buttons.
- Click the Indent or Unindent buttons.
- Select an alignment (Left, Center, or Right) from the Alignment drop-down list.

Insert a Hyperlink ➤ pp. 210-212

1. Select text.
2. Click the Link button or choose Insert | Link.

Make an Internal (Local) Link ➤ pp. 211-212

1. Click the Link icon.
2. If you know the name of the local file you want to link to and it's in the same folder (directory) as the document you're working on, just type the filename directly into the Link to a page location or local file box. Otherwise, click the Choose File button.
3. Browse around and find the file you want and then click Open.
4. Click OK.

Make an External (Remote) Link ➤ pp. 213-215

1. Click the link icon.
2. Type or paste the URL directly into the Link to a page location or local file box.
3. Click OK.

Make a Graphics Image into a Hyperlink ➤ p. 216

1. Select the image.
2. Click the Link button or choose Insert | Link.
3. Follow the steps outlined above for making an internal or external link.

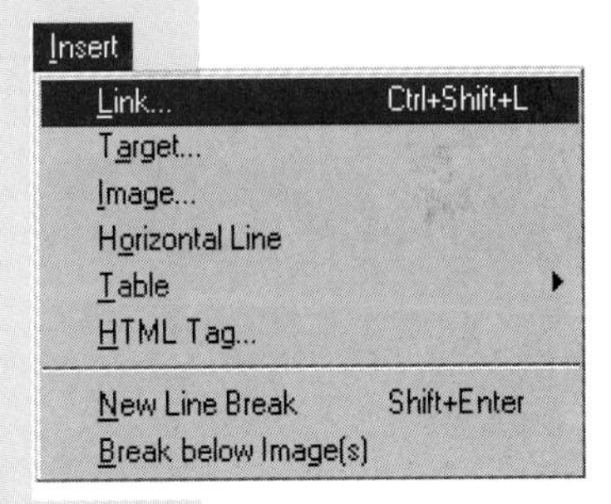

If you've ever used a word processor or even a text editor, then Netscape Composer should be very easy for you to learn. HTML documents lack many of the elements of paper documents you create on a word processor, so you don't have to worry about things like page numbers. (It would be nice if Netscape Composer had a Search and Replace feature, though!)

It works just like a word processor: you see what you're typing in the main window, and you give commands by choosing them from menus or by clicking shortcut buttons.

Is This for Me?

In a word, *yes*—if you're planning to create or edit any Web page. This chapter covers fundamental tools and formatting features—stuff you'll need to know.

Netscape Composer Toolbars

Most of the useful editing and formatting commands are available on the two toolbars at the top of the window, the Composition toolbar (see Table 10.1) and the Paragraph toolbar (see Table 10.2). You can tell what most of the buttons do by their names.

New	Create a new Web page
Open	Open an existing page
Save	Save the current page
Publish	Publish the page to the Web
Preview	Preview the current page in Navigator
Cut	Cut selected text or graphics
Copy	Copy selected text or graphics
Paste	Paste items from clipboard
Print	Print the current page
Find	Find text in the page
Link	Insert a hyperlink
Target	Create a hyperlink target

Table 10.1 The Composition Toolbar

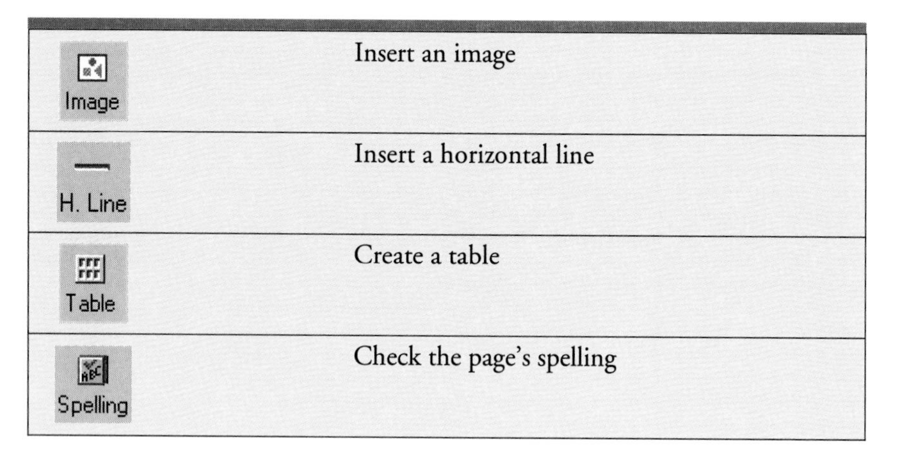

Image	Insert an image
H. Line	Insert a horizontal line
Table	Create a table
Spelling	Check the page's spelling

Table 10.1 The Composition Toolbar (*continued*)

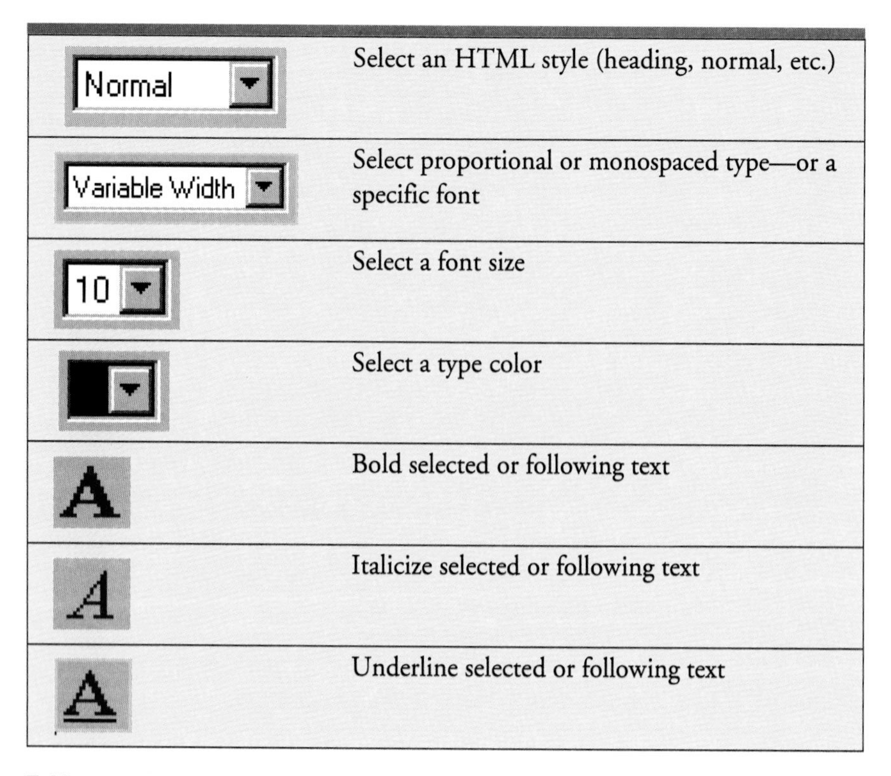

Normal	Select an HTML style (heading, normal, etc.)
Variable Width	Select proportional or monospaced type—or a specific font
10	Select a font size
	Select a type color
A	Bold selected or following text
A	Italicize selected or following text
A	Underline selected or following text

Table 10.2 The Paragraph Toolbar

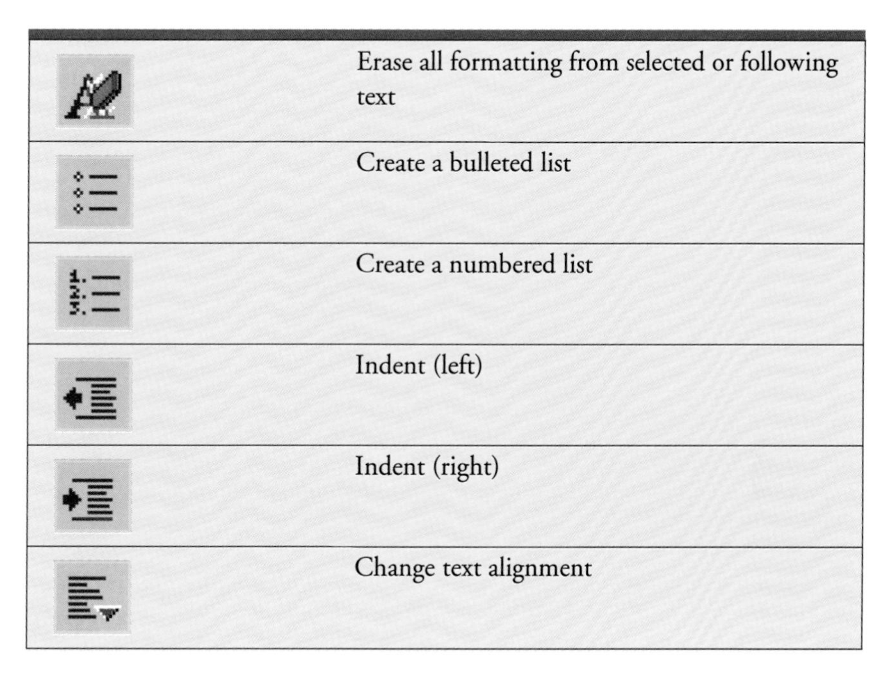

	Erase all formatting from selected or following text
	Create a bulleted list
	Create a numbered list
	Indent (left)
	Indent (right)
	Change text alignment

Table 10.2 The Paragraph Toolbar (*continued*)

Giving Your Document a Title

Give your document a title before you save it for the first time. You'll feel a satisfying sense of accomplishment when the name of your page appears in the title bar.

To give your document a title, select Format | Page Colors and Properties to bring up the Page Properties dialog box, then click the General tab (see Figure 10.1).

Type a new title in the title box and then click OK. Netscape Composer will display as much of the new title as possible in the title bar, along with the name of the source file (which, since you haven't saved it yet, is still file:///Untitled in Netscape's parlance).

When Bad Things Happen to Craz : file:///Untitled - Netscape Composer

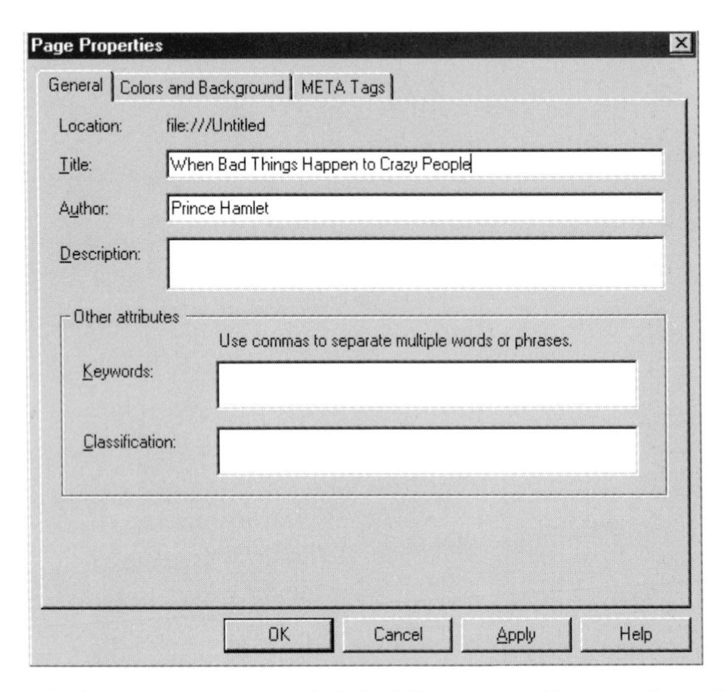

Figure 10.1 Type your document's title in the Title text box of the Page Properties dialog box

CAUTION

Without a title on your page, you will look like a rube on the Web, on people's bookmark menus, and in search engines, when you're listed at http://www.yerprovider.com/~yername/jimiweb.html (or what have you).

Typing Text in Your Page

To begin with, just start typing as you would in a word processing document. Netscape Composer handles word-wrapping for you. To start a new paragraph, press ENTER.

Move your insertion point the usual ways, either by pressing an arrow key or by clicking your mouse where you want the insertion point to appear. As your document gets longer, you can use the scroll bars to move around in it.

Save Regularly!

As soon as you've typed anything of consequence at all, save your document. Save it just about every time you change it. Saving is more important than flossing! Think back to the last time you lost a *lot* of work by not saving. Don't let it happen again.

Saving for the First Time and Thereafter

The first time you save your document, you have to give it a name. Keep it simple. It's best to plan out ahead of time what the names of the various files are going to be.

The home page of a complicated set of pages is often called index.html, because many Web servers use index.html as their default file name. This can save your readers some typing when they visit—it's easier to type http://foo.com than http://foo.com/company_ home_page.html.

EXPERT ADVICE

Since Netscape Navigator guesses not just the http:// but also www. and .com, it's a good idea to pick the single (as yet untrademarked) word for your business and just give that out as the URL. For example, ***shareware.com, download.com****, and* ***coffeehousebook.com*** *have done just that.*

To save and name your document, click the Save button or choose File | Save. Choose the folder you want to save your page in. Then type a file name and click Save.

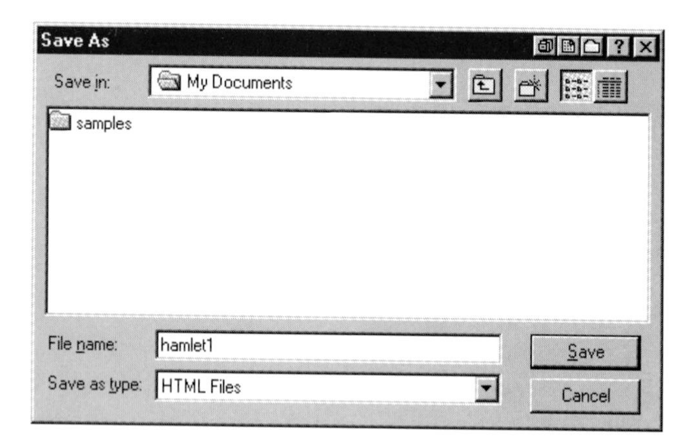

After you've saved the document once, you can just click the Save button to save additional changes as you go. Remember, saving is not the same as publishing. Saving updates the local copy on your computer. Publishing updates the Web server.

Editing Your Document

As you type or revise a document, you inevitably find mistakes, make corrections, change things, and so on. And yes, you do these things in Netscape Composer much the same way you would in a word processor. You can select text (or lines or images) by clicking and dragging. To delete what you selected, just press the DELETE key. To cut a selection so you can move it elsewhere, click the Cut button (or right-click the selection and select Cut).

To copy a selection instead of moving it, click the Copy button (or right-click and select Copy), move the insertion point where you want to copy the selection, and then click the Paste button (or right-click and select Paste).

Formatting for Structure

OK. The boring preliminaries are out of the way. You can start formatting your Web document! Just as with magazines, people don't want to read a Web page filled with gray expanses of text. Instead, they're looking for well-organized information that's broken into digestible chunks. Use headings to grab attention and clearly mark the different parts of your document. Insert line breaks and horizontal rules to space text, keep it organized, and make it easy to read.

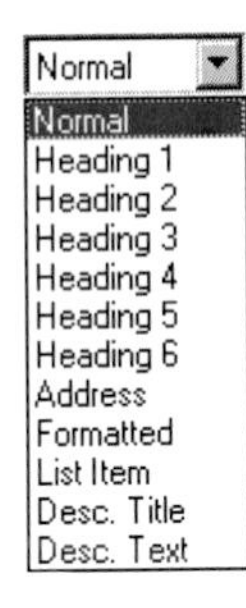

Making Headings

It's easy to make headings with the Paragraph Style drop-down list box on the Paragraph Properties toolbar.

Just select the text you want to make into a heading (or put the insertion point where you want to start typing your new heading), click the Paragraph Style drop-down list, and select the heading level you want. After you type the heading and press ENTER, the next line will be a regular paragraph.

Line Breaks

Remember, you're making a new paragraph every time you press ENTER. If you want to start a new line and stay in the same paragraph (in most browsers, this prevents skipping a line), just press SHIFT-ENTER. This inserts a line break and is commonly referred to as "breaking" the line of text. A line break is not the beginning of a new paragraph.

Horizontal Rules (Lines)

Another way to structure a document is to insert horizontal lines (rules) to separate different sections. To insert a horizontal line, just click the H. Line button.

Figure 10.2 shows a horizontal rule. It separates an introductory section and list of contents from the first item on a page.

You can also customize the thickness, width, and appearance of horizontal rules. To do so, select a rule (click anywhere on it) and then select Format | Horizontal Line Properties (or just right-click on the rule—click and hold with a Macintosh—and then choose Horizontal Line Properties from the pop-up menu). This brings up the Horizontal Line Properties dialog box:

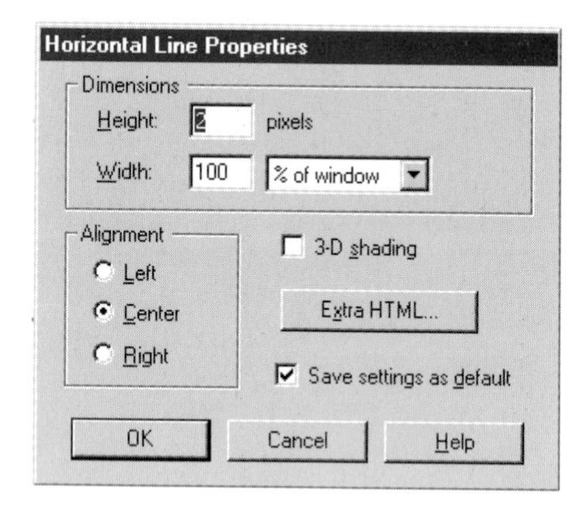

- To change the thickness of the line from the default of 2 pixels, type a different number of pixels in the Height box.
- To change the width of the line from the default of 100 percent of the screen width, first choose either % of window or pixels and then type a percentage or number of pixels for the width of the line.

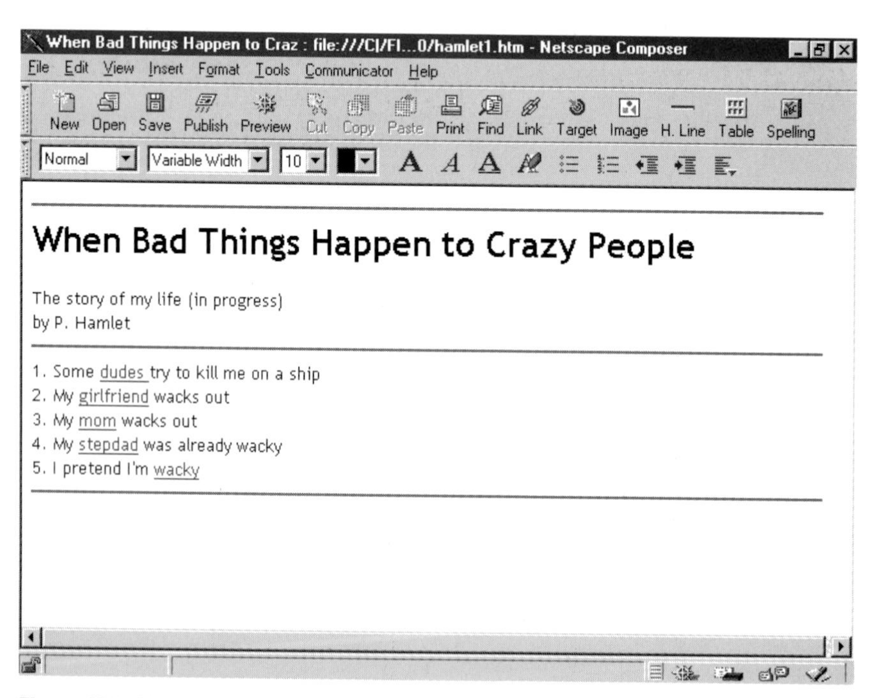

Figure 10.2 Use horizontal lines (rules) to separate sections of text

DEFINITION

Pixel: *One dot on the computer screen; the smallest unit the screen can display.*

- To change the alignment of the line, choose Left, Center, or Right.
- To eliminate the 3-D "beveled" appearance of the line, uncheck 3-D shading.

Formatting for Design and Clarity

Of course, you'll format text for more than organization—you'll want to make it look good. Netscape Composer makes it very easy to format text in different ways, such as selecting text and clicking a button or choosing an option from a list or dialog box. You can format your text a character, word, or sentence—or a whole paragraph—at a time. Let's give both ways a look. Then you'll learn how to format lists.

Character Formatting

You can specify how you want any character or group of characters to appear.

EXPERT ADVICE

Technically, HTML allows for two types of character formatting, logical and physical tags. Logical tags, such as "emphasis," can be interpreted differently in individual browsers. Netscape Composer doesn't support these (except for "address.") Instead, it lets you use physical formatting, which tells browsers literally how to display the text. Italics mean italics, bold means bold, and so on.

Bold, Italics, and Underlining

The most popular forms of character formatting are available from the toolbars—bold, italic, and underlining.

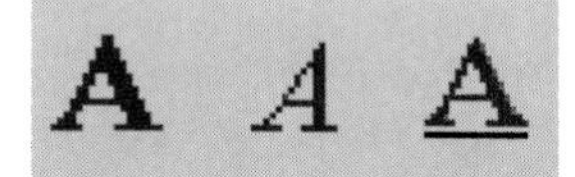

To make existing text bold, for example, select the text and then click the Bold button. Or, if you're typing along and want to type in boldface, just click the Bold button to turn boldface on, type the text you want in bold, and click the button again to turn it off.

EXPERT ADVICE

Sometimes fixed-width *text is considered character formatting—it's monospaced text, where each character and space take up the same amount of room on the line. You can choose this from the second drop-down menu on the formatting toolbar.*

Superscript, Subscript, Strikethrough, and Blink

These occasionally useful types of character formatting are also easy to apply: just select the text you want to format and choose Format | Style. Then select the formatting style you want from the submenu that appears:

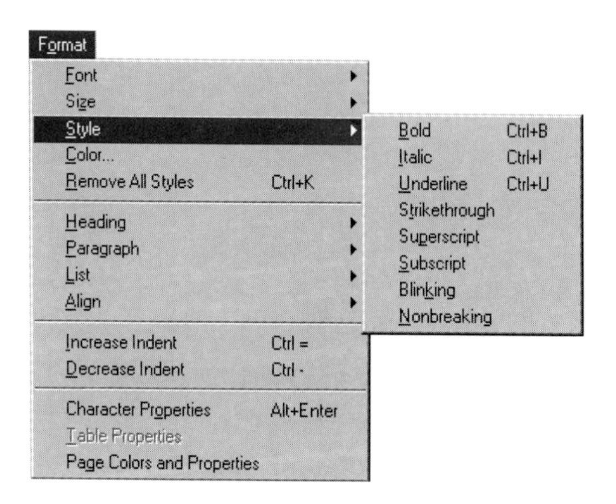

For the record:

- Superscript places text above the baseline and reduces its size (as in "$e=mc^2$").
- Subscript places text below the baseline and reduces its size (as in "H_2O").
- Strikethrough (used most often to indicate deleted text) marks text with a horizontal line (as in "The offer is \$100~~,000~~.00").
- Blinking makes text blink on and off.

Font Size

To change a font's size, select some text (or place the insertion point where you intend to type the new size of text), then click the Font Size drop-down list and choose from a range font sizes.

Font Color

You can change the color of any selected text (or text you're about to type) by clicking the Font Color button, which brings up the font color palette:

Click the color you want from the palette.

One-Stop Character Formatting

If you want to apply more than one type of formatting to a selection, right-click (click and hold with a Mac) and choose Character Properties (or select Format | Character Properties).

This brings up the Character Properties dialog box with its Character tab selected (Figure 10.3).

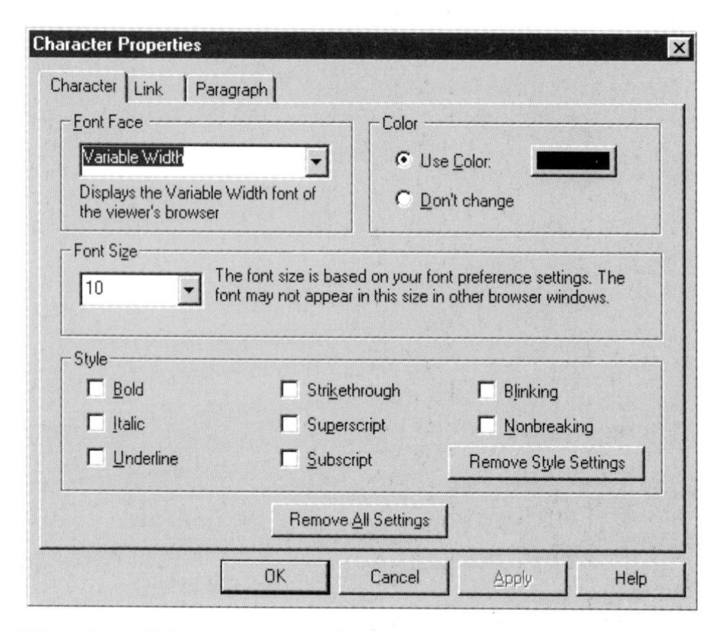

Figure 10.3 Select all the character attributes you like using the Character Properties settings—from fixed-width typeface to bold and italic

Choose as many forms of character formatting as you feel like piling up on your selection. (You can, for example, combine bold and italics.) Change the color or size if it suits you. When you're done, click OK.

Removing Character Formatting

If you change your mind about some character formatting you've applied, select the text again and then click the Remove Styles button.

Paragraph Formatting

There are a few formatting effects you can apply to entire paragraphs (as well as to headings and other kinds of text). Among them are indentation, alignment, block quotes, and addresses.

Physical Indentation

One type of paragraph formatting is indentation. To use this effect, select the text you want to indent (or unindent), or position the insertion point at the new indented (or unindented) paragraph you want to type, and then click one of the indentation buttons.

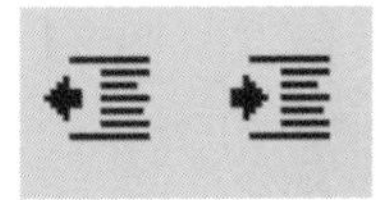

Left, Center, and Right Alignment

More useful are the alignment commands. Although they're not supported by all browsers, they'll allow you to horizontally align paragraphs along the center axis or the left or right margins. Select the text you want to align and then click a button.

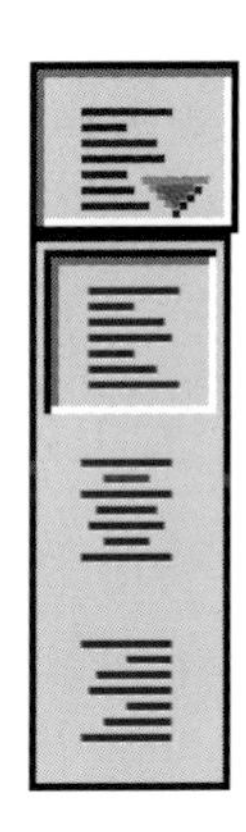

Block Quotations

If you are quoting more than a couple of lines from somewhere, it's customary to set the quotation off from the main text in its own paragraph, typically indented from either margin. To apply this kind of formatting, right-click the paragraph(s) and choose Paragraph | List Formatting from the pop-up menu (or select Format | Character Properties, then the Paragraph tab).

This brings up the Character Properties dialog box with the Paragraph tab selected (see Figure 10.4).

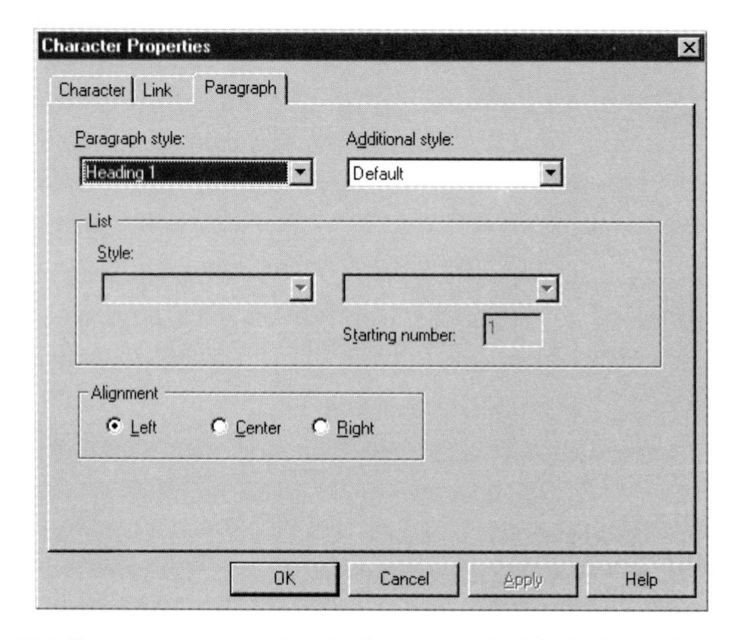

Figure 10.4 Format your paragraphs with the paragraph tab in the Properties dialog box

Click the Additional style drop-down list and choose Block Quote, then click OK.

Formatting Lists

Lists are one of the most common ways of organizing information in a Web document. They are well suited for communicating various and sundry facts, instructions, and ideas.

Netscape Composer makes it pretty easy to create the most popular types of lists: ordered (numbered) lists and unordered (bulleted) lists. You can also put one list inside another to create what are called "nested lists." And, if you're so inclined, you can fool with the numbering scheme of lists and even get bullets of different shapes and sizes.

Unordered (Bullet) Lists vs. Ordered (Numbered) Lists

A bulleted list is one in which each item begins with a filled-in black circle called a *bullet.* You don't have to type the bullet symbols to make a bulleted list, because the Web browser (or Netscape Composer) does that for you. A numbered list numbers each item sequentially—and Netscape Composer does that automatically, too.

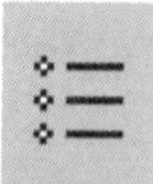

To start typing a bullet list, click the Bullet List button.

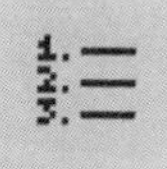

To start typing a numbered list, click the Numbered List button.

Netscape inserts the first bullet or number. Type the first item in the list and press ENTER when you are done. Netscape inserts a bullet or number at the start of the second line. Type the second item on the list, press ENTER, and repeat as often as necessary.

CAUTION

It's easy to be confused when you're creating a numbered list, because Netscape Composer doesn't insert the actual number—instead, it represents the number with a number sign (#). When the page appears in a browser, it will replace the # with the appropriate numbers.

When you get to the end of the list, press ENTER one more time. Netscape inserts one more bullet or number than you want, but don't worry about that. Just click the List button again to "unpush" it. The last bullet or number disappears, and the line that the cursor is on returns to normal paragraph formatting.

You can also create a list by typing the items as you would regular paragraphs. When you're done writing the list, select all the paragraphs you wrote and click the appropriate List button. This technique works just as well.

Other Kinds of Lists

Netscape Composer offers three other types of lists that you, frankly, will probably never create: definition lists, directory lists, and menu lists. Most browsers display the last two the same way they display unordered lists (as single-column bulleted lists). Definition lists, which Netscape Composer calls "description" lists, are used more frequently than the others, but they appear differently from browser to browser. You can format text in these list formats by choosing Format | List, then choosing the list type you want from the pop-up menu.

Previewing Your Page Formatting

When you want to see how your current document will look in the Netscape browser window, save it. Then click the Preview button.

Netscape will open a new browser window and open your current document in it so that you can see how it will appear when it's viewed on the Web.

It Ain't a Web Page Without Links

Look: if you've used the Web for more than two minutes, you probably already know what a hyperlink is. Click a highlighted—possibly underlined—word, and you're taken to a new page or section within the current page.

But you didn't really ask for anything, at least not by name, did you? No, you just clicked on (or selected) a link of some kind. The browser did the rest. We'll show you how to insert such hyperlinks into your document with Netscape Composer.

Links Have Two Parts

There are two ends to each hyperlink (referred to, technically, as *anchors*). At one end is the hyperlink (text or graphics) button. At the other end is the link's target. The target can be within the same document, on the same Web server, or anywhere on the Web.

Sensible Linking

In planning out the hyperlinked organization of your Web site, you have to give your visitors some reasonably understandable routes or pathways through your site. You also want to make sure that they can easily get back to the home page from anywhere within the site. People browsing the Net can generally use their Back buttons to retrace their steps, but if they jumped directly into somewhere in the middle of your site, they won't have any easy way of returning to a preceding page. Similarly, as people follow links deeper and deeper into your site, you'll want to make sure they can easily work their way back to the "surface" whenever they feel the need to.

One useful application of hyperlinking is to include a table of contents at the top of a long document, usually consisting of the major headings in the document. Each of the headings in the table of contents can then be a hyperlink to the section it names, giving the reader an easy way to jump directly to the part of the document he or she is interested in.

How to Actually Do It

With Netscape Composer, inserting hyperlinks is simply a matter of selecting text and then clicking the Link button or dragging a link (or a local document) directly into the Web document you're working on. Write your first draft, edit it, and format it until it looks good, and *then* put the links in. (You can still change things later.)

There are two broad categories of links, by the way, usually referred to as internal or external links. *Internal* ("relative") links connect among pages (or objects) located on the same computer (server). *External* ("absolute") links connect to pages (or objects) located somewhere else on the Net.

Select the Text for the Link

No matter what kind of link you're about to insert, you usually do so by first selecting text from which you want to link. With formal, organized links, such as links from each entry in a table of contents to the referred articles, the text to choose is obvious. With "in context" links, you'll have some choice about what text to select and treat as a link.

If you like, you can also simply position the insertion point with no text selected, and then go ahead to the next step. You'll be able to type in text for your link when you select the document you're linking to.

Internal, or Relative, Hyperlinks

Once you've selected text, click the Link button.

You'll then see the Character Properties dialog box with the Link tab selected (see Figure 10.5).

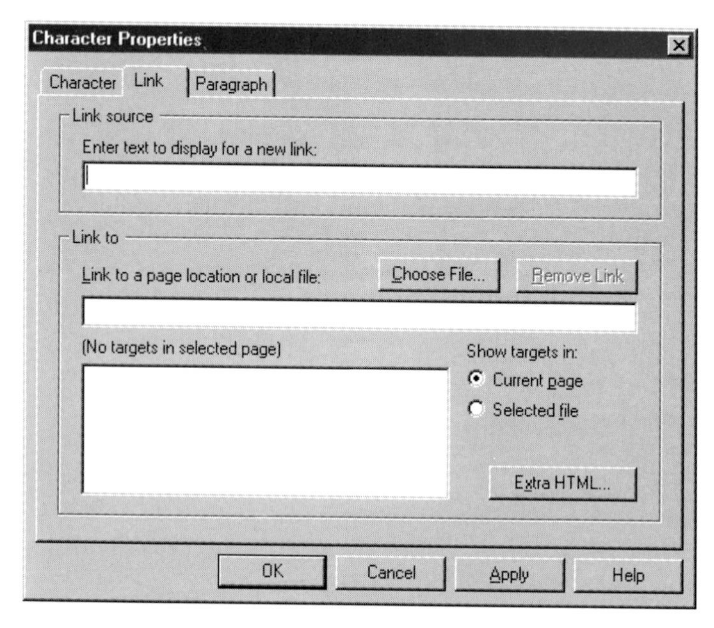

Figure 10.5 Use the Character Properties dialog box to create hyperlinks

If you know the name of the local file you want to link to and if it's in the same folder (directory) as the document you're working on, you can just type the filename directly into the Link to a page location or local file box. Otherwise, click the Choose File button.

This brings up the Link to File dialog box, a dead-ringer for the normal Open dialog box. Browse around and find the file you want, and then click Open. The path leading to the file you want will tell the browser how to find it *relative* to the location of the current document.

Click OK. The text you selected will appear in your preferred link color (or perhaps in your "visited link" color, if you've displayed the linked document recently).

Inserting Named Anchors

Earlier, we mentioned the idea of named anchors, which are marked sections of a document to which you can link directly. Just as you must create a document before you can link to it (or at least before you can find it by browsing to insert a link to it), you must also create any names before you can link directly to them.

To insert a named anchor into a document, first select the text you want to name (think of it as a hyperlink *target*). Then click the Target (Named Anchor) button.

This brings up the Target Properties dialog box with a suggested name for the anchor (the selected text itself, or at least the first 255 characters of it). Shorten the name or change it to something more memorable, if you want, and then click OK. Netscape Composer inserts a Target icon at the beginning of the named anchor to remind you that it's there.

Linking to a Named Anchor

Of course, you can link from within a page to any named targets. You can even link to other pages with named targets. To do so, select the text and click the Link button to bring up the Link tab of the Character Properties dialog box. The Select a named target in current page box will display any names in the document you're editing.

To link to a name in the current document, click one of the listed names. It will appear (preceded by a #) in the Link to a page location or local file box.

To link to a named anchor in a local document, first click the Choose File button and select the document you want, as explained in the previous section. Any names in that document will then appear in the named target list, and the radio button to its right will move from the Current page position to the Selected file choice.

Click the name you want. It will be added to the filename (separated by a # character, as in www.ton.com/kilo.htm#ounce).

External, or Absolute, Hyperlinks

Most Web sites have some links to other sites or pages on the Web. External links are essentially the same as internal links, except the reference to the linked page has to be a complete URL, rather than just a filename or a relative path and a filename.

A URL Refresher Course Fortunately, most of the links you'll be inserting into your pages will already exist in some electronic form (in e-mail people have sent to you, in your bookmarks file, or on other people's Web pages), so you hardly ever *have to* type URLs, because you'll have them handy to copy and paste from elsewhere.

EXPERT ADVICE

As a courtesy, you may want to contact the Webmasters of sites to which you want to link. It's not required, but it's a nice thing to do, especially if you're linking to somewhere other than the home page.

URLs start with a protocol, which, nine times out of ten, is http:// (other likely protocols are ftp://, gopher://, mailto:, and telnet://). Next comes the actual Internet address of the site hosting the document you're referring to. It's usually two or three words separated by dots, either hostname.subdomain.domain or just subdomain.domain. After that, somewhat rarely, there can be a colon and a port number (such as :1080). A URL can end right there, if you want to refer to the primary document at a Web site (the root, or home page of the site). Or there can be a forward slash followed by either a path (if the document you're referring to is not in the root directory of the Web server), a path and a filename (if the file you're looking for is not the default one at the folder referred to in the path), or just a filename.

But these are all just technicalities you need to have straight when you're typing in these long URLs. Once they're embedded into documents (yours or others), it becomes a matter of clicking links. The true power of the Web is that you *can* embed a reference to just about any type of resource out there on the Net, and your browser will try to haul back whatever you've linked to for you.

Typing in External Links Directly You still start by selecting the text you want to use as the hyperlink. Then click the Link button (or right-click—or hold down the mouse button, for Macs—and select Create Link Using Selected). This brings up the Character Properties dialog box with the Link tab selected, as shown in Figure 10.5.

Now type or paste the URL directly into the Link to a page location or local file box. Then click OK.

Linking to a Named Anchor To link to a named anchor (target) within a document at an external site, the process is almost the same. You still select the text you want to use as a link and bring up the Character Properties dialog box with the Link tab selected. You still enter or paste in the address of the document in question.

Then, to specify a name, type a pound sign (#) after the address and the text of the name itself, then click OK.

Dragging Links onto Your Page If you *don't* have the exact text of a URL handy at your fingertips, it's easy enough to drag it from the Web itself (or from any page that contains a link to the address you want). To do so, first switch to Netscape Navigator (or select Communicator | Navigator). Then find your way to the document you want to link to. If you get to the exact page you want, first reduce the size of the Navigator window so that you can also see part of the Editor window showing the portion of the document where you want to insert the link. Then click the icon to the left of the Location box and hold down the mouse button.

Drag over to the Editor window. You'll see the mouse pointer turn into a special Insert Link icon.

Position the pointer where you want the new link to appear and then release the mouse button. The link will appear in the document, with link text that's the title of the page you've linked to or just the URL of any other kind of file. You can change either one.

Copying Links from Any Web Document Another way to insert a link into the document you're working on is to copy it from an existing page. To do so, just right-click on an existing link and select Copy Link Location.

Then return to the document you're editing, position the insertion point, and choose the Paste button.

Editing Link Text

If you want to overrule Netscape's suggestion for the link text, just click at the beginning of the link (place the insertion point just *after* the first letter of the link text), type new text, and *then* delete the original text, including the first letter.

Removing a Link

If you ever create a link by mistake or in the wrong place, you can remove it easily. Just right-click on the link (yes, or with a Mac, click and hold down), and then select Remove Link. There's also a Remove Link command on the Link tab of the Character Properties dialog box, if you happen to have it open already.

Images as Hyperlinks

As mentioned earlier in this chapter, graphic images can also be turned into hyperlinks. (The process is the same, except you must select the graphic first, instead of text, before assigning the link.) Select the image you want to use as a link, then choose Insert | Link. You'll see the Image dialog box with the Link tab selected—it looks just like the Link tab shown in Figure 10.5.

You'll probably want to make a few changes to just how the image appears. If you already have the Image Properties dialog box on your screen (with the Link tab displayed), you can just click the Image tab. Or, right-click the image and choose Image Properties. You'll see this dialog box:

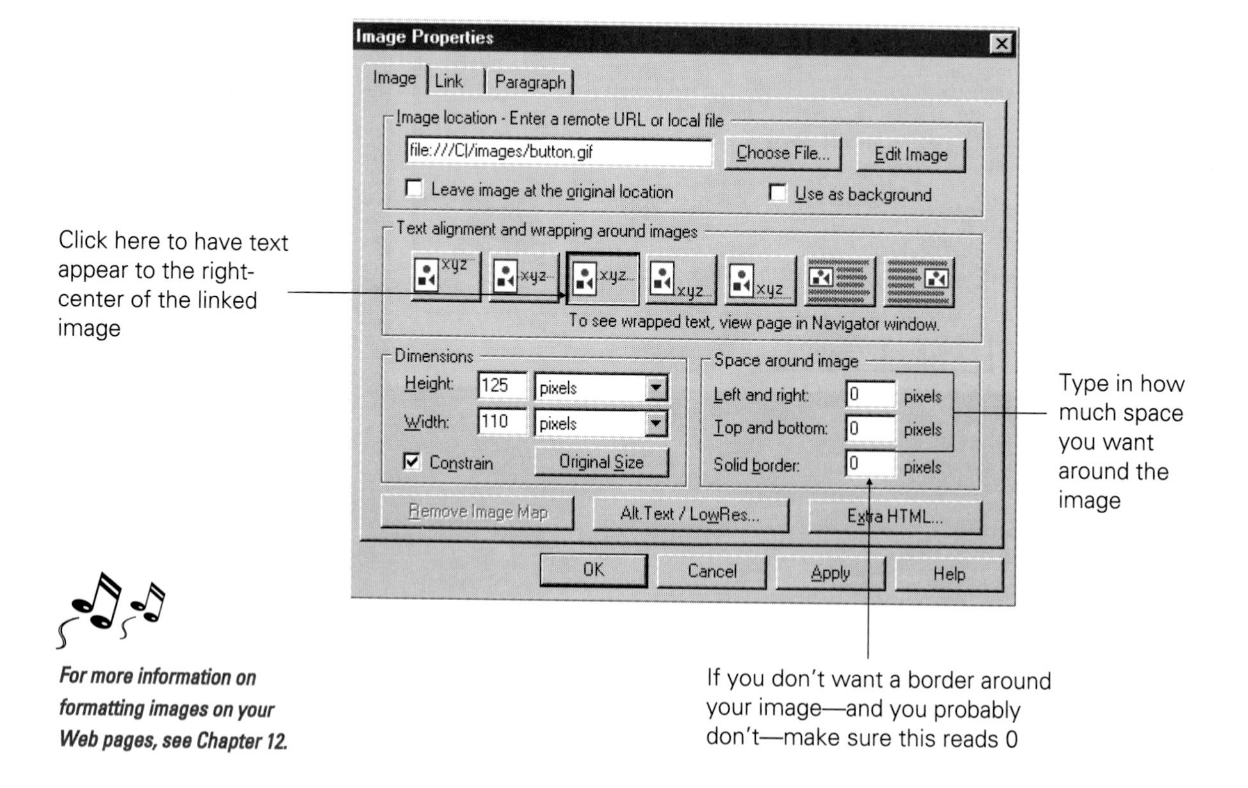

For more information on formatting images on your Web pages, see Chapter 12.

In the next chapter, you'll learn some advanced Web page design tricks, from page backgrounds to fancy layouts using tables.

CAUTION

CHAPTER

11

Advanced Web Page Design with Colors, Tables, and More

INCLUDES

- Choosing page colors and backgrounds
- Creating tables with Netscape Composer
- Entering text in tables
- Editing tables
- Formatting tables
- Converting text to tables

FAST FORWARD

Choose Colors and a Background Image for Your Page ➤ pp. 223-227

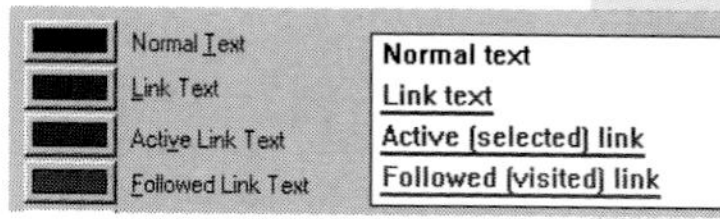

1. Select Format | Page Colors and Properties, and click the Colors and Background tab.
2. Choose the colors for the text elements you want.
3. Select a background image, if you want one.
4. Choose OK.

What Is a Table? ➤ p. 227

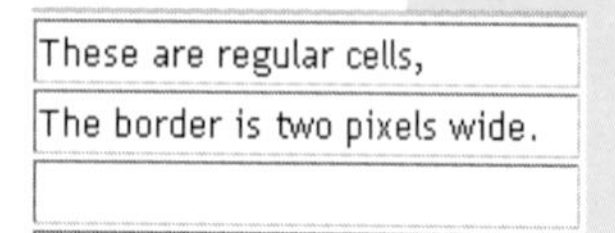

- A *table* is a section of a document divided into rows and columns.
- The intersection of each row and column is called a *cell*.

Insert a Table into Your Document ➤ pp. 227-228

1. Click the Table button.
2. Type the number of rows you want in your table.
3. Press TAB.
4. Type the number of columns you want.

Format Your Table as You Insert It ➤ pp. 228-231

1. Press TAB to jump to the Border line width box.
2. Type a width for your border and press TAB.
3. Type a number of pixels for spacing between cells and press TAB.
4. Type a number of pixels for spacing within cells.
5. Check Table width.
6. Select Pixels or % of Window.
7. Type a number of pixels or percentage (if not 100).
8. Check Table min. height, if you wish to set one (or skip the next steps if you don't).
9. Select Pixels or % of Window.
10. Type a number of pixels or percentage (if not 100).

Enter Text in Your Table ➤ *pp. 232-233*

1. Position the insertion point in the first cell.
2. Type the contents of that cell.
3. Click in the next cell over.
4. Type the contents of that cell. Repeat as often as necessary.

- When you are in the last cell of a row, pressing RIGHT ARROW will jump you to the first cell of the next row.
- If you are in the last cell of the last row, pressing RIGHT ARROW will jump you out of the table.
- To move back to the previous cell, press LEFT ARROW.

Insert Rows, Columns, and Cells ➤ *pp. 233-234*

- To insert a row, position the insertion point in the row *above* where you want to insert the new row. Then select Insert | Table | Row.
- To insert a column, position the insertion point in the column *to the left of* where you want to insert the new column. Then select Insert | Table | Column.
- To insert a cell, position the insertion point in the cell *to the left of* where you want to insert the new cell. Then select Insert | Table | Cell.

Delete Rows, Columns, and Cells ➤ *p. 235*

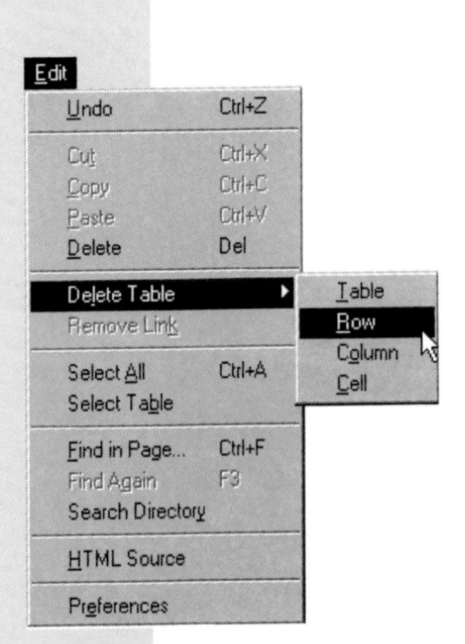

- To delete a row, position the insertion point in the row and select Edit | Delete Table | Row.
- To delete a column, position the insertion point in the column and select Edit | Delete Table | Column.
- To delete a cell, position the insertion point in the cell and select Edit | Delete Table | Cell.

In this chapter, you'll learn how to add some fancier design elements to your Web pages. You'll learn about customizing text and background colors and adding custom background images. You'll also learn the basics of tables—the building blocks of advanced page design.

Is This for Me?

While these features are nonessential, they're useful for giving your pages a polished look. If you're creating only a basic page, you can skip this chapter. But these techniques can make the most basic pages appear professional.

Customizing Your Page's Look

In the beginning, Web pages consisted of black text on a gunmetal gray background. Since then, pages—and HTML—have grown to sport custom text and background colors, plus graphical "wallpaper" files as background art.

Of course, if you don't specify any of these custom design elements, your pages will still appear in black text on gray.

EXPERT ADVICE

Any user can, of course, customize their own browser and choose any color scheme they want, but very few users do this. Most accept the default color scheme and thereby any color scheme imposed on pages by their designers.

Some Basic Design Principles

Just because you *can* play around with the color scheme doesn't mean you *should*. Consider it a design option, but make sure, first and foremost, that your pages are readable and usable. For example, maintain a sharp contrast between text and background colors. Used well, color selections can add a professional, custom appearance to your pages.

EXPERT ADVICE

Stick generally to dark text colors and pale, muted background colors. If you plan to use a graphical image as a background, you'll likewise be better off with a pale or faint image rather than one which will compete visually with the overlying text.

Choosing Colors for a Page

Netscape Composer offers you two ways of customizing the color scheme of a Web page: choose one of its predefined color schemes, or select specific colors for each of five elements (text, links, visited links, active links, and background).

No matter which method you plan to use, you start by selecting Format | Page Colors and Properties. This brings up the Page Properties dialog box. Click the Colors and Background tab (Figure 11.1).

For a new, blank document, Use viewer's browser colors will be selected. This means the page will appear using whatever color settings each page viewer's browser has set. (Most people leave Netscape's default color scheme in place.) But, since you're reading this, you probably want to customize your page's colors—so click Use custom colors.

Using a Predefined Color Scheme

As when using Martha Stewart's coordinated color palettes when you decorate your home, you can save a lot of time using Netscape's predefined color schemes. Try it: click the Color Schemes drop-down menu.

Scroll through the list using your arrow keys. As you scroll through the list, the preview area will reflect the selected scheme.

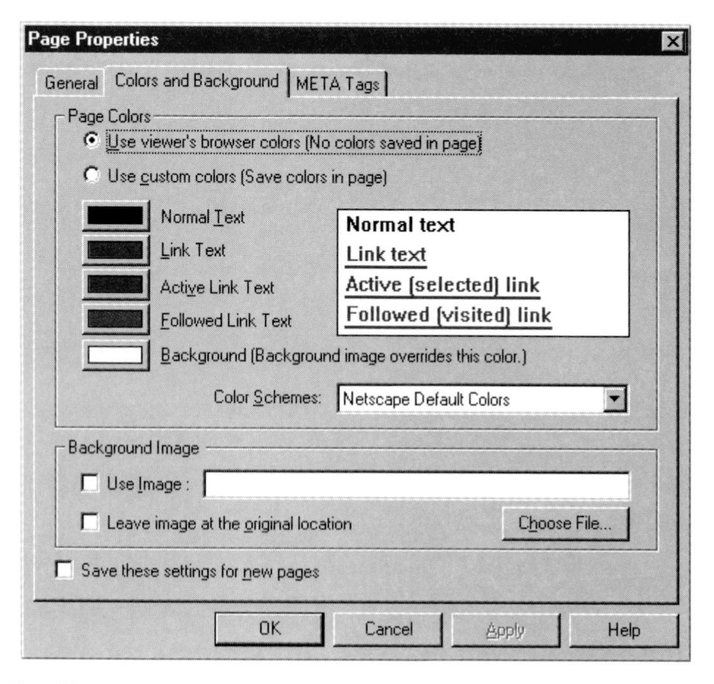

Figure 11.1 Choose a predefined color scheme or customize each page element individually

If you find a color scheme you like, click OK. If not, try choosing custom colors for each of the design elements.

Selecting Text Colors

You'll choose colors for each of the four text types in the same way. To select a basic color for nonhyperlink text, click the Normal Text button in the Custom colors area.

This brings up the Color palette. It offers a basic palette of 48 colors—just click the one you want.

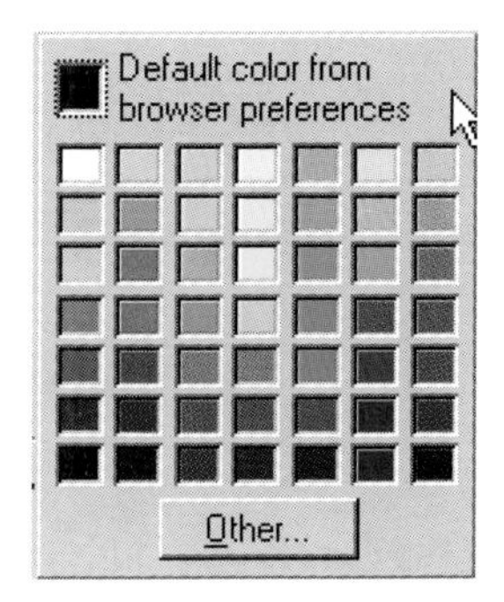

Repeat the process for each type of text—Normal text, Link text, Active Link text, and Followed Link text.

Background Colors

Select a background color the same way you chose text colors. Just click the Background button, then click the color you want from the palette.

Defining Your Own Colors

If you're not satisfied with any of the default 48 colors, choose your own. Click the button for the text element or background, then click Other on the palette. You'll see the Color dialog box (Figure 11.2).

The square area plots hue (what most people call color, as in red, green, blue, etc.) versus saturation (which you can think of as how bright or dull the color is). The vertical bar next to it allows you to choose luminosity (which is the amount of lightness or darkness in the color). But you don't have to think in such quasi-scientific terms. Really, you just click in the square to get roughly the right color you want, and then click in the bar to fine-tune it. You'll probably have to go back and forth a little to get it right.

When (and if) you find a color you like, click the Add to Custom Colors button to add it to one of the blank swatches in the Custom colors area in the bottom left of the dialog box. This new color will now be part of the Other color palette and can be selected for any of the colored elements.

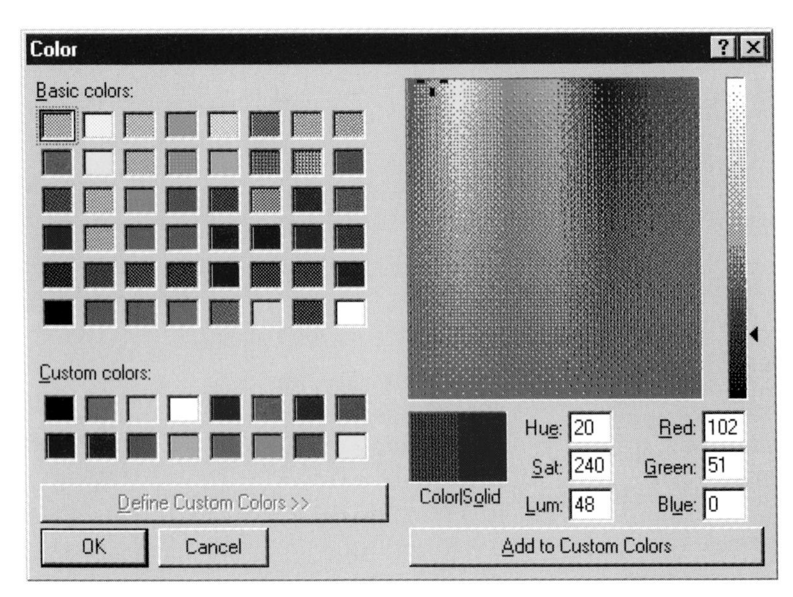

Figure 11.2 If you don't like one of the default colors, you can choose almost any color you like from this dialog box

Adding Background Patterns to Your Pages

Many pages don't use just a color as a background image. They use well-chosen, attractive graphics images designed to tile, that is, to lie side by side to fill the screen. If you have an image you want to use as a background tile, insert its path and file name here in the Use Image box of the Page Colors and Background dialog box (Figure 11.1). Not every image can work as a tile, because the edges have to fit together without looking awkward or jarring. (To create your own, try using Kai's Power Tools with Adobe Photoshop.)

You can find tiles for free on the Internet, although you can't necessarily use them for commercial purposes without permission or without paying for them. You can also buy CD-ROMs full of background tiles and other stock images.

You can create a parchment effect for your backdrop by scanning a piece of slightly rough or off-white paper and cropping the scan down to a small, nondescript portion.

Another useful background trick is to create a long, narrow, two-tone tile and use it to create a strip of color across the top or left side of the page.

Graphic designers hate the Web. Why is that? Because it was invented to share academic and scientific information and to standardize documents so that computers can index them easily. This emphasis on content and intention over presentation and cosmetics makes it difficult to design a page layout anywhere near the level of sophistication you might find in your typical print magazine page.

That's why tables were regarded as such a godsend when they were proposed in HTML 3.0 and implemented in Netscape and then later in other browsers. While tables have their own specific purposes, namely the presentation of tabular information (because information is easier to grasp if organized and related spatially), tables can also be used to divide up the page (screen) visually, to simulate margins, side-by-side columns, and more. The Web might never be a complex, page layout–oriented medium, but tables go a long way toward giving you, as a designer, more control over the positioning of the text and graphics on the page.

Page Design with Tables

A *table* is a section of a document divided into rows and columns. Each intersection of a row and column is called a *cell*. A cell can contain text, hypertext, graphics, or anything you'd put in a Web page—or it can be empty.

You can adjust the appearance and size of a table's borders, column width, and captions (see Figure 11.3).

Notice that we are not really dwelling on the tags needed to create tables, since Netscape Composer automates the process for you (and it's a good thing, too—tables are about the most complicated things you can code by hand!).

Making Tables

To insert a table into a Web document, click the Table button (or select Insert | Table | Table).

Table

This brings up the New Table Properties dialog box (see Figure 11.4). There are a lot of options, but we'll walk you through.

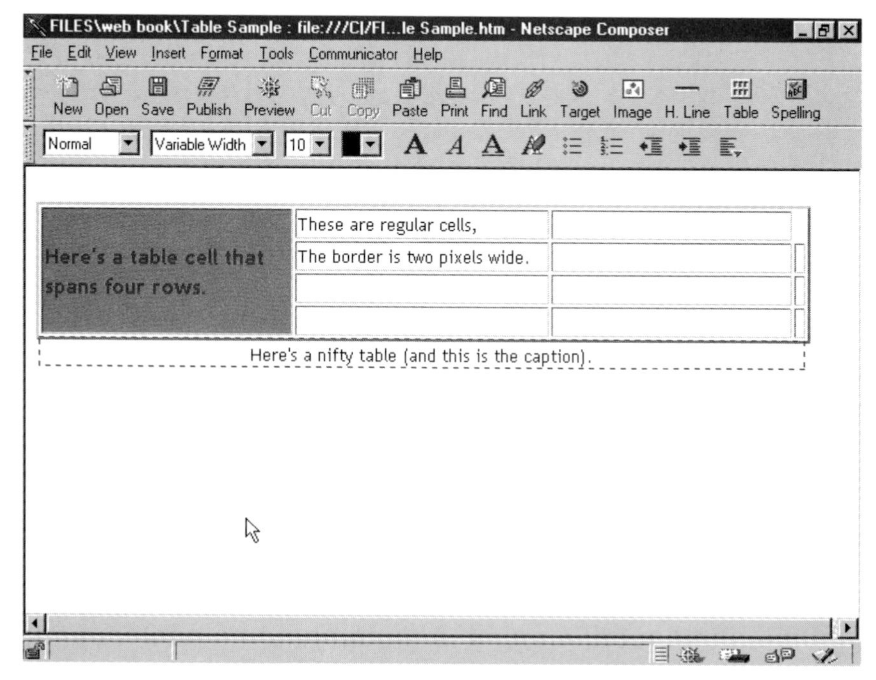

Figure 11.3 Adjust everything you want to make your Web table look snappy

Rows and Columns

First of all, you have to settle on a number of rows and columns. You can change your settings later, but make a generous estimate—it's easy to lop them off if you don't need them.

1. Type the number of rows you want to start off with.
2. Press TAB.
3. Type the number of columns you want to start off with.

Borders, Spacing, and Padding

Next, you have to think about the spaces around and between the table's cells. There are three regions you can control: the border around the outside of the table, the spacing between cells, and the padding between the text region inside a cell and its boundary:

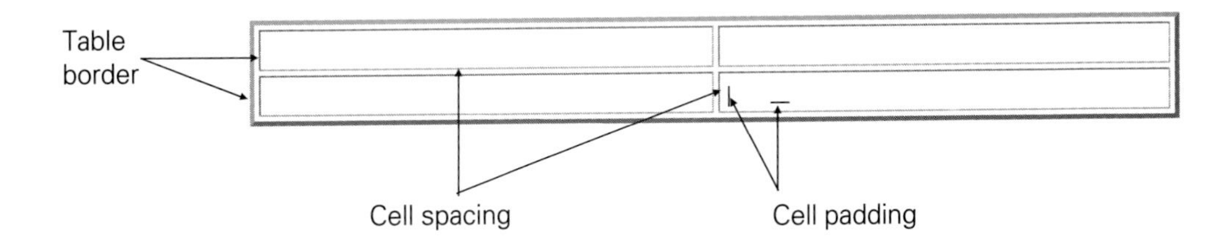

The border is the line around the outside of the entire table (keep it small). Cell spacing helps you organize your table layout. Cell padding makes text easier to read—it prevents it from bumping up against the sides of the cells.

1. Press TAB to jump to the Border line width box.
2. Type a width for your border and press TAB.

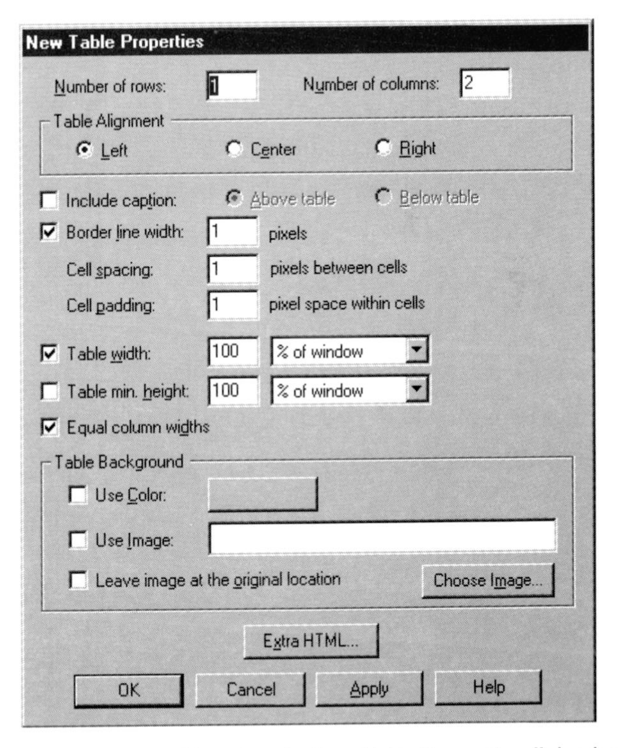

Figure 11.4 Customize your table using the New Table Properties dialog box

3. Type a number of pixels for spacing between cells and press TAB.
4. Type a number of pixels for spacing within cells.

EXPERT ADVICE

Sometimes the smoothest effect will be achieved with borders of zero, which means no beveled lines between cells or around the outside of the table. This effect can be used to create sophisticated, magazine-style layouts.

Table Width (and Height)

You don't have to set the table's width, but you'll probably want to. If you leave it unspecified, the columns' widths will be determined by the text in the cells. You can set the table width to a specific number of pixels or to a percentage of the width of the screen. The advantage of specifying the width in pixels is that you'll know exactly how wide the table will appear, and you can use this to line up text with graphics or other elements with a fixed width. But, unless you're a professional designer, the disadvantages to using pixels outweigh the advantages. Unless you know details about your reader's screen (size, resolution, font size), you'll have no idea how the page is going to look, and you may make a table too wide for some screens.

The advantage to setting a table's width in terms of a percentage is that you'll have some idea of how the table will look relative to any screen width. The disadvantage is that the contents of the table will wrap and fill the cells differently depending on the details of the reader's screen setup, which may destroy any visual effects you were aiming to achieve.

Minimum table height can be set the same way table width is set. But it's much less useful, unless you want to make sure that your table fills the entire screen, for example, or some set percentage of it.

1. Check Table width.
2. Select Pixels or Percent of Window.
3. Type a number of pixels or percentage (if not 100).

4. Check Table min. height, if you wish to set one (or skip the next steps if you don't).
5. Select Pixels or Percent of Window.
6. Type a number of pixels or percentage (if not 100).

Customized Cell Colors

Although it's conceptually a bit of a non sequitur, you can set a color for the background of the cells in the table (it can be different from the background color on the rest of the page). Figure 11.5 shows a table with cells a different color from that of the page.

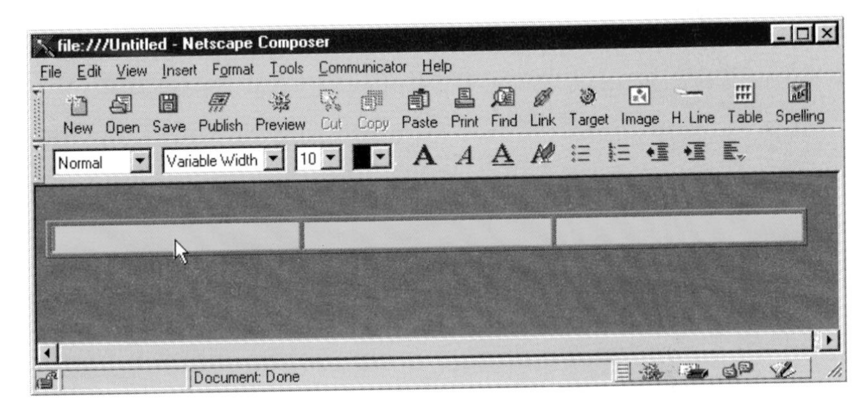

Figure 11.5 You can set your cell backgrounds to a different color than the page background, hopefully with a better combination than this one

1. Click the Choose button.
2. Choose a color from the color dialog box that pops up.
3. Click OK.

Attaching a Caption

A caption is a bit of text attached to the table that usually describes or introduces it (yes, like a caption under a photo in a magazine). You can format the text of a caption just as you would normal text.

1. Click Include caption if you want a caption for your table.

2. Click Below table if you want the caption to appear below the table instead of above it.

Choosing Table Alignment

If the table width is less than the full width of the screen, choose Left, Center, or Right alignment. (This is a good way to simulate left and right margins.)

Finishing Your Table

When you've made your way through the dialog box, click OK. Your table will appear.

If it doesn't look exactly the way you pictured it, remember that there's no text in it and you haven't typed a caption yet.

Entering Text in Tables

To begin entering text in your table, position the insertion point in the first cell and just start typing. The cell will expand to accommodate the text you type. When you are done with the first cell, click in the next cell over, or press the RIGHT ARROW key to jump there.

Type the contents of that cell. Repeat as often as necessary. When you are in the last cell of a row, pressing RIGHT ARROW will jump you to the first cell of the next row. If you are in the last cell of the last row, pressing RIGHT ARROW will jump you out of the table. To move back to the previous cell, press LEFT ARROW. To move up one row to the cell immediately above the current cell, press UP ARROW. To move down one row, press DOWN ARROW.

Besides typing the contents of the table, you can also type your caption above or below it. The caption has no special formatting associated with it, so, besides typing it, you should format it as you deem appropriate (make it bold, for example, or center it, and so on).

EXPERT ADVICE

If you're in the last cell on the last row, pressing TAB adds a new row to the end of the table.

You can cut, copy, and paste text to and from table cells, but a selection cannot cross a cell border. Therefore, you can't, for example, copy the contents of two cells from one place to another at the same time. You have to move the content piecemeal. You can only cut, copy, and paste cell contents, not the actual cell, row, or table structure.

Editing Your Table

Once you've entered the text in your table, you'll inevitably want to make changes. The way you edit text in a table is not terribly different from how you do it in a normal Web document. You can make selections within cells and delete them, or cut, copy, or paste them. If you make a selection that crosses a cell boundary and then try to delete or cut the selection, Netscape Composer warns you that this is not permitted:

You may also want to change the very structure of your table. As we mentioned before, when you first create your table, you're not locking yourself into the exact dimensions you initially pick. As you edit and sometimes rethink table content, you might eventually want to change the number of rows, columns, or even cells in a specific row. Netscape Composer makes it pretty easy to insert or delete any table element.

Inserting

To insert a row, position the insertion point in the row *above* where you want to insert the new row. Then select Insert | Table | Row. Netscape Composer inserts a new blank row below the row containing the insertion point. The new row has the same number of cells with the same formatting as the row above. Now type the text for the new row.

You can't directly insert a new row at the top of a table. To work around this limitation, put the insertion point in the top row and insert two rows (one after another). Type the contents for your intended top row in the first of the two new rows. Then copy, cell by cell, the contents of the old first row into the second new row. Finally, delete the old top row, as explained in the next section.

To insert a column, position the insertion point in the column *to the left of* where you want to insert the new column. Then select Insert | Table | Column. Netscape Composer inserts a new blank column to the right of the column containing the insertion point. The new column has the same number of cells with the same formatting as the column to the left. Now type the text for the new column.

As with rows, you can't directly insert a new column at the left side of a table. Put the insertion point in the leftmost column and insert two columns (one after another). Type the contents for your intended left column in the first of the two new columns. Then copy, cell by cell, the contents of the (old) first column into the (new) second column. Finally, delete the old leftmost column, as explained in the next section.

Because each row is allowed to contain a different number of cells, you can also insert a cell into any row. This will generally necessitate the creation of a new column containing only a single cell. To insert a cell, position the insertion point in the cell *to the left of* where you want to insert the new cell. Then select Insert | Table | Cell. Netscape Composer inserts a new blank cell to the right of the cell containing the insertion point and pushes each cell to the right of the new cell over one cell to the right. Now you can type the text for the new cell.

It's also possible to insert an entire table into a cell of another table. This enables you to get really creative with your layout because you can align text in columns and rows *within* a bigger arrangement of cells.

To insert a table into an existing table, put the insertion point in the cell where you want the new table to appear, and click the Insert Table button or select Insert | Table | Table. This brings up the New Table Properties dialog box shown in Figure 11.4. Follow the instructions in "Making Tables," earlier in this chapter, to complete the inserted table. Figure 11.6 shows a partial Web page that uses a table within a table.

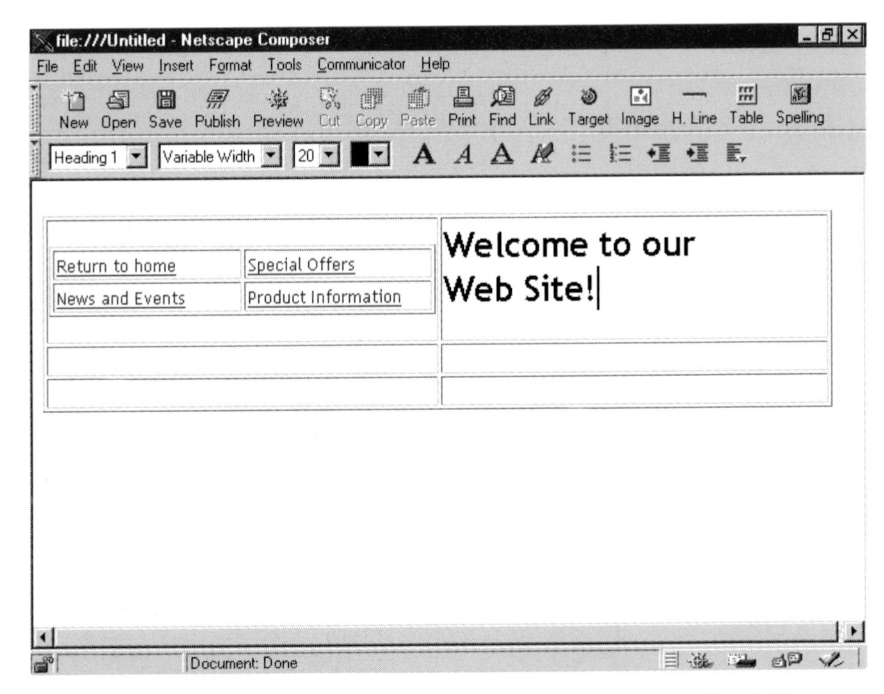

Figure 11.6 This Web page in progress shows a possible use for a table within a table

Deleting

To delete a row, position the insertion point in the row and select Edit | Delete Table | Row. Netscape Composer deletes the row. If you change your mind, select Edit | Undo.

To delete a column, position the insertion point in the column and select Edit | Delete Table | Column. Netscape Composer deletes the column.

To delete a cell, position the insertion point in the cell and select Edit | Delete Table | Cell. Netscape Composer deletes the cell and moves any cells to its right, one position to the left.

To delete an entire table, or a table within a table, put the insertion point in the table you want to delete and then select Edit | Delete Table | Table. The table is gone.

Moving Rows and Cells

There's no way to cut and paste entire rows or columns or multiple cells at one time. The only way to achieve the goal of moving a portion of a table is to insert a new row, column, or cells, as described in the earlier section on inserting table elements, cut and paste the contents of each cell, one at a time, and then delete the vacated row, column, or cell. Repeat this as often as necessary.

Formatting Your Table

As with editing, there are two aspects to formatting a table—formatting the text (in the traditional sense), and formatting the elements of the table itself (the rows, columns, and cells). If all you want is to make some of the text in a table bold, for instance, you can just select the text and click the Bold button, as you would normally. (For the purposes of applying formatting, you *can* select the contents of more than one cell at a time.)

To format the table itself, you just drop by the Table Properties dialog box. First, place the insertion point in the specific cell, row, or table you want to affect. Then, either select Format | Table Properties or right-click on the table (on the Mac, click and hold down the mouse button) and choose Table Properties from the menu that pops up.

The Table Properties dialog divides its options up into three tabs: Table, Row, and Cell. The Table tab simply restates the original formatting options you saw when you created the table, so we'll leave that till last.

Formatting a Row

To control alignment or color in a single row, click the Row tab of the Table Properties dialog box (see Figure 11.7).

Alignment There are two kinds of alignment in a row: horizontal and vertical. The default horizontal alignment is left. The default vertical alignment is center. If you have a specific position for your text, you should specify it and not rely on defaults, which might produce different results in different browsers. Most of the time, for example, you'll probably want Top vertical alignment instead of the default.

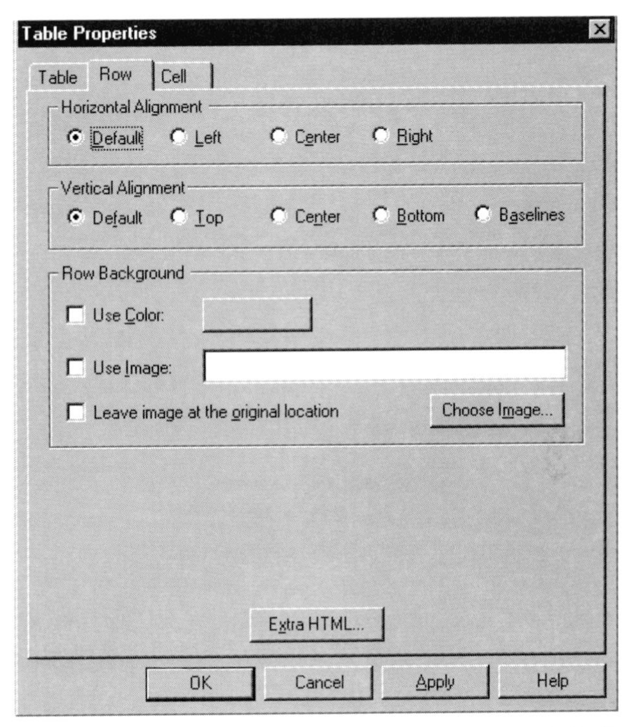

Figure 11.7 Adjust your row's appearance with the Row tab of the Table Properties dialog box

Color To specify a different background color for an entire row, check the Use Color box, then click the button to the right.. The Color dialog box will appear. Choose a color—or create a custom color. Then click OK.

Formatting a Single Cell

To control alignment or color in a single cell, click the Cell tab of the Table Properties dialog box (see Figure 11.8).

Spanning Cells One neat trick you can do with tables is create cells that span one or more columns or rows (or both). Cells made to span multiple columns or rows "expand" to the right or down, respectively. Here's a table with a header cell spanning three columns:

This cell spans all three c

This cell spans all three columns below.

Cells can span rows as well.

Alignment Cells have all the same alignment options as rows. This means that you can set general alignment for a row and then change the specific alignment for a particular cell (or cells).

Headers and Wrapping There are two commands lumped together in the poorly named Text Style area of the dialog box: Header Style and Nonbreaking. By default, both are turned off (unless the row you're in was created as a header).

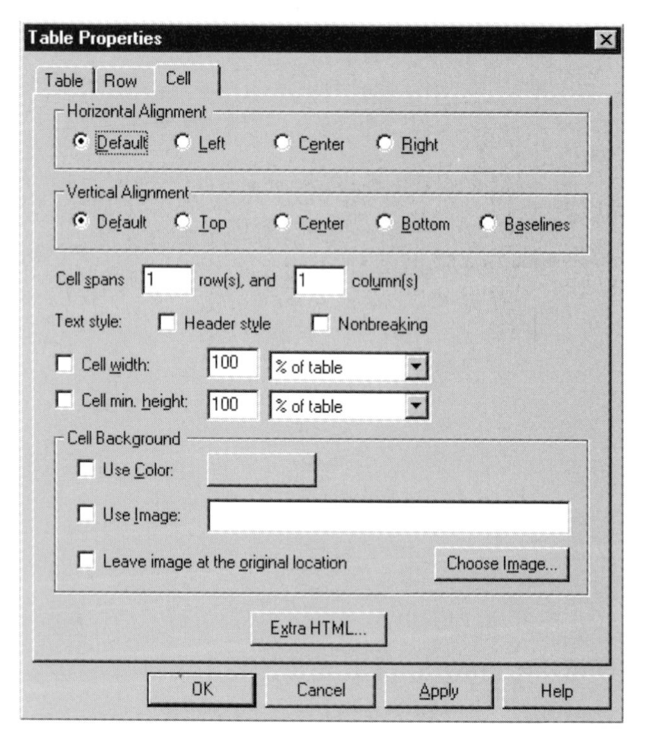

Figure 11.8 Adjust a cell's appearance using the Cell tab in the Table Properties dialog box

Checking Header Style just turns the current cell or row into a header, which is mainly a conceptual distinction (although it may center the header text, in some browsers).

Checking Nonbreaking will cause the text in affected cells to stay on a single line, even if that means being cut off by the limited size of a cell.

Width You may notice, as you enter text into tables, that the columns adjust themselves for some sort of "best fit" based on the current text as entered. This is fine for an ad hoc table, but not so good if you're trying to lay out the page with some regular column widths or proportions. That's why the width option is the most important one in this dialog box (if you're using tables to create a page layout, that is).

To fix the width of selected cells, first check Cell width, then either choose Pixels or keep the % of Table option in the drop-down list box. Then, enter a number of pixels or a percentage of the table width (if not 100).

Repeat this process on as many cells as necessary to fix the proportions of the table.

Minimum Height If you need a selected cell to have a specific height, either in terms of absolute number of pixels or as a percentage of the table's height, you can check Cell min. height. Then choose Pixels or % of Table, and then enter a number (if not 100).

Color As with all the other table dialog boxes, the Cell tab of the Table Properties dialog box allows you to select a background cell color for the selected cells. Click the Choose Color button, choose a color from the Color dialog box that pops up, and then click OK.

Rethinking Overall Table Format

To reconsider the overall formatting of the table (such as the choices you made when you first created the table), click the Table tab on the Table Properties dialog box. The Table tab recaps all the options available when you first created the table, except for the numbers of rows and columns.

Converting Between Text and Tables

Unlike many word processing programs, Netscape Composer can't automatically convert between a table format and regular text format. Instead, if you have text in some plain style (or organized as preformatted—what Netscape calls Formatted—text), you have to create the table in a blank space first, and then cut and paste the text elements into the table by hand. Likewise, to convert table text back into some straightforward layout, you have to "rescue" all the text from the table, one cell at a time, and then delete the table when you're done.

EXPERT ADVICE

If you need to convert text to tables or vice versa, it's a good idea to use word processors such as Microsoft Word or Corel WordPerfect.

What makes the Web so much more exciting than almost every other aspect of the Internet? The pictures, of course. In the next chapter, you'll learn to add more visual interest to your Web pages with graphics, multimedia, and more.

CHAPTER

12

Adding Zip: Graphics, Multimedia, Forms, and More

INCLUDES

- Graphics formats that work on the Web
- Making and converting graphics
- Inserting graphic images into Web pages
- Multimedia: right for your audience?
- Image Maps, Frames, and Forms

Graphics Formats That Work on the Web ➤ *pp. 245-246*

There are only two graphics formats widely viewable on the Web:

- GIF, which is generally better for solid-color images with limited palettes, *and*
- JPEG, which is generally better for color photographs.

Convert Existing Graphics ➤ *p. 247*

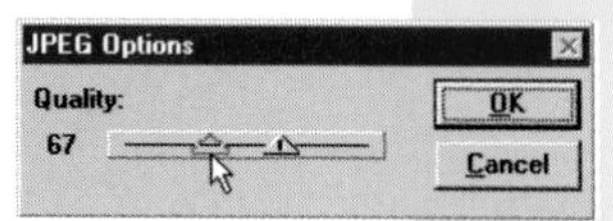

1. Some scanner software can save images as both GIFs and JPEGs.
2. If you have a program that can convert between GIFs and JPEGs, all you need is to be able to get existing images into one of those formats and you'll have both options.

Insert Graphics into Pages ➤ *pp. 247-248*

1. Before inserting a graphic, place the insertion point where you want it to appear.
2. Click the Insert Image button.
3. Type the path and filename of the image, if you know it, or click the Choose File button to the right of the Image filename box, find the file you want in the Choose Image File dialog box, and then click the Open button.
4. Click OK.

Provide Alternate Text for Images ➤ *pp. 248-249*

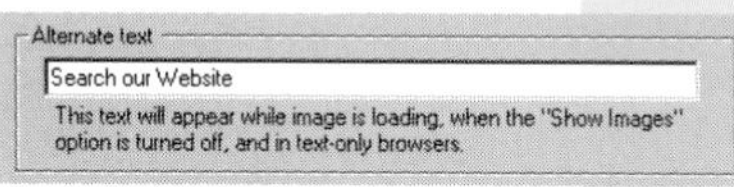

While still in the Image Properties dialog box, after specifying an image file:

1. Click the Alt. Text/LowRes button.
2. In the Alternate Text box, type some text to explain the image to those who can't see it (or type "" if you want nothing to display in place of the image).
3. Click OK twice.

Be Thoughtful to Your Audience ➤ p. 252

- Use multimedia (sound, animation, etc.) sparingly, if at all.
- Use extra media elements only when they serve a definite purpose.

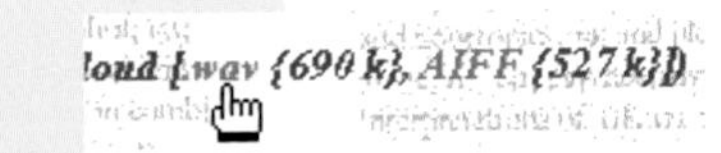

Build a Form ➤ pp. 252-254

1. Start with <FORM METHOD=GET *or* PUT ACTION="*URL of script for form*">.
2. Put buttons in your form with <INPUT TYPE="submit" VALUE="*optional text, instead of Submit*"> and <INPUT TYPE="reset" VALUE="*optional text, instead of Reset*">.
3. To make the form itself, use the following tags:
   - <TEXTAREA NAME="*name*" ROWS=*x* COLS=*y*>*optional suggested default text*</TEXTAREA>
   - <INPUT NAME="*name*" TYPE="*text, password, checkbox, radio, int, hidden, submit, reset*" VALUE="*default value, if any*" SIZE="*optional size*" MAXLENGTH="*optional maximum length*">
   - <SELECT NAME="*name*" SIZE="*optional number of items to be shown at a time*"> ... </SELECT>
4. End with </FORM>.
5. Find and install a script to make the form work (that's the catch), or find someone to do it for you.

What do you like about our products?

☐ They're cool ☐ They're cheap

Submit Query Reset

Three Ways to Create Image Maps ➤ pp. 256-258

- For traditional, server-side maps, create a map reference file and post it to your Web server.
- For a new-wave, client-side image map, use the <MAP> and </MAP> tags to designate hot regions on the map (within the HTML coding of the document containing the map) and add a USEMAP="*name*" command to the <IMG> tag for the map graphic.
- For a "fake" image map, cut your image up into a series of graphics and insert them each as borderless hyperlinks, one right next to the other.

Since 1994, a Web site has had to include graphics to look "state of the art." How many magazines do you know that are all text and no pictures? Graphics on pages are a big part of what makes the Web so popular, humanizing the previously cold, text-oriented Internet.

Common types of graphics include custom logos, headlines and banners, buttons (that is, clickable image links), scanned photos, pie charts, and graphs. You can use graphical elements to create a visual look and identity for your pages. You can also use them for illustration or visual explanations.

As important as graphics are to the overall effect of a well-designed Web page, you should also make sure that your pages will look and function adequately when viewed as text only. Some of your audience may be using text-only Web browsers. Even if you expect the majority of your readers to use graphical browsers such as Netscape Navigator, a lot of people do their browsing, at least initially, with automatic image loading turned off. (Also, some Web users are blind and have text-only browsers that read or "speak" the content out loud.)

As a matter of course, you can and should assign alternate text to each image so that a text-only user will get the gist of its content.

Don't Try to Do It All Yourself

One of the difficulties of designing a new Web site is that one typically aims for the sort of results that have, traditionally, taken a team of professionals. The two areas where you are most likely to benefit from calling in outside expertise are graphic design and the maintenance of the network or Web server (and, hence, your site).

The best sites have just the right amount of emphasis on both appearance and functionality and are usually created by well-coordinated teams.

Web Graphics Formats

Although this may change, currently there are only two graphics formats *widely* viewable on the Web: GIF and JPEG.

Generally, you can assume that if a Web browser can display any graphics at all, it can display GIFs and JPEGs. If you are adapting existing art, you'll need to convert it to one of those two formats. How do you decide which to use? Here are the pros and cons.

Pros and Cons of GIFs

GIFs can handle 256 colors, which is adequate for most graphical-illustration purposes. The amount of compression you can achieve with a GIF depends on the level of detail in your image, but you can't adjust it. GIFs don't handle photographs with subtle shading gradients very well. Figure 12.1 shows an image that compresses well as a GIF.

Figure 12.1 This image is well-suited for use as a GIF

There are two tricks you can do with GIFs that you can't do (yet) with JPEGs. The first is transparency. Most programs that can create GIFs can assign one of the colors in the palette to be transparent. This enables you to create art that appears not to have square edges.

GIFs can also be saved in an "interlaced" format, which means that a savvy browser (such as Navigator or Internet Explorer) can immediately display a rough image after just a little bit of it has been downloaded and then sketch in the details as they come in. GIFs can also be animated—basically, they're mega-GIFs that consist of several smaller images that are shown as if in a slide show, one after another.

Pros and Cons of JPEGs

JPEG is the superior format for photographs; in fact, its standard was arrived at for that specific purpose. JPEGs can store up to millions of colors. When creating a JPEG (saving a file or converting it from another format), you can choose a degree of compression, keeping in mind that greater compression means greater loss of the original image data (see Figure 12.2).

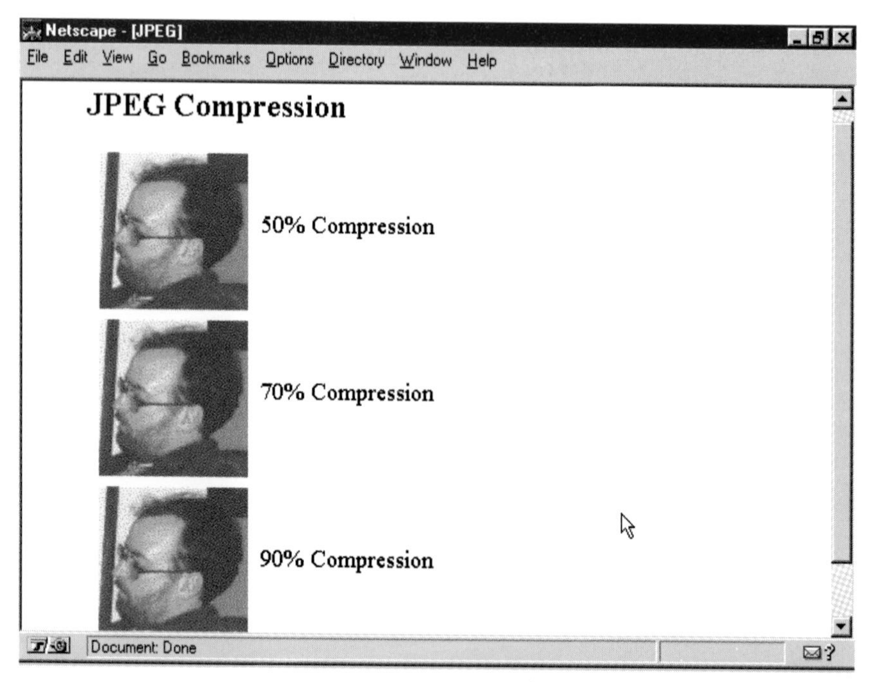

Figure 12.2 JPEG images at three levels of compression

Creating Graphics

There are two ways to create art for your Web page. You can "draw" it using a drafting and art program and create it from scratch with the computer. Or you can scan existing photography or line art and either manipulate it to your specific needs or reproduce it as close to the original as possible.

Obtaining Graphics

There are many ways to find graphics if you can't create (or commission) your own. Clip art and stock images are available on CD-ROM, generally with royalty-free, use-anywhere licenses. There are also images on the Net that are free for both download and use. Make sure you have permission before you use an image.

Converting Existing Graphics

Most of the time, you'll need to convert your graphics to GIF or JPEG format. Some scanner software can produce one or both of these formats automatically. If you have a program that can convert between GIFs and JPEGs, then you just need to be able to get existing images into one of those formats to have both options.

Adobe Photoshop is the most widely used tool for professional-strength image manipulation, but other products, from the high-end Corel DRAW to the inexpensive shareware Paint Shop Pro, can also do the trick nicely.

Inserting Graphics into Your Pages

Once you have your graphics prepared for the Web, inserting them into your pages is a breeze with Netscape Composer.

Formatting Graphics While Inserting Them

Before inserting a graphic, place the insertion point where you want it to appear and follow these steps to the easiest way to insert a graphic into your document:

1. Click the Insert Image button. The Properties dialog box will appear (see Figure 12.3).

2. Type the path and filename of the image, if you know it, or click the Choose File button to the right of the Image filename box. Find the file you want in the Choose Image File dialog box and then click the Open button.

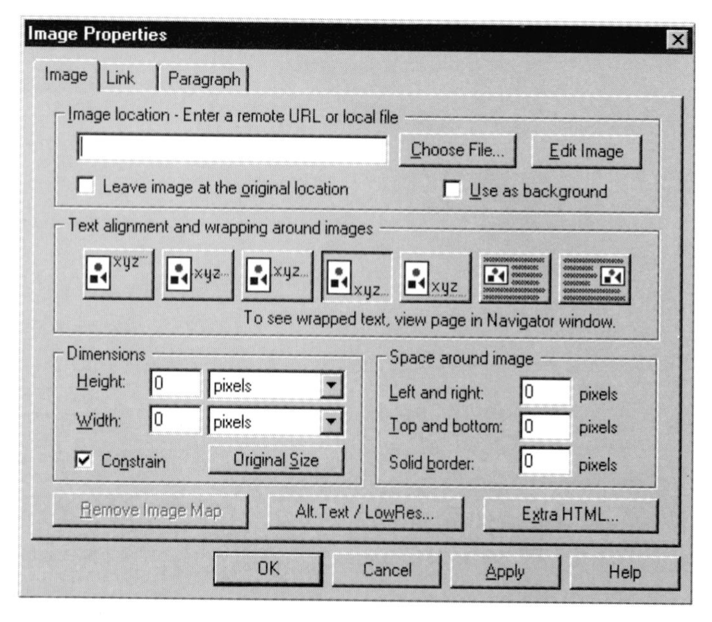

Figure 12.3 Control how your image will appear using the Image Properties dialog box

Essential: Alternate Text

If you want to insert alternate text—to display while the graphic is loading or to explain the image to those who can't see it—click the Alt. Text/LowRes button. Type some text in the Alternate Text box, or type "" if you want nothing to display in place of the image.

Figure 12.4 shows a Web page with image-loading turned off and alternate text showing in place of images.

To make it possible for text-only browsers to download an image file, you have to make it into a link to itself.

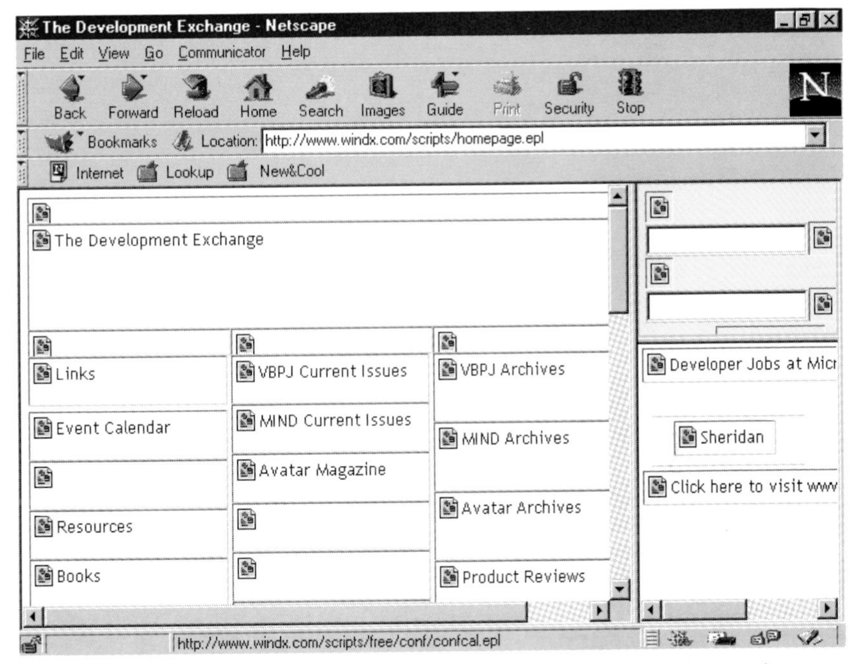

Figure 12.4 Alternate text is essential. As images load, you can still discern the page structure

Low-Resolution Alternate Image

If you want a low-resolution alternate image to display first to give your reader something to look at before the real art arrives, press TAB and type the path and filename of the other image file, or click the Choose File button. Find the file you want in the Select Low Resolution Image dialog box and then click the Open button.

Here's a low-resolution graphic being updated by a higher-resolution image:

Alignment

Now it's time to decide how you want the image to fit in with the text on the page. Most of the time you'll want text to appear below the image or alongside it. Netscape Composer offers all the standard alignment choices, presenting them in the form of pictorial buttons that demonstrate how the text will wrap around the graphic:

If you want text to appear next to your graphic, you'll want to put in some horizontal spacing to put "breathing room" around the picture. How to do that is coming up.

Specifying Height and Width

Any time you specify an image's height and width in your document, you help it load faster in a browser. The browser can leave the right amount of space for the picture and keep laying out the page (namely, the text that has to flow around the images) without waiting for the image to arrive before continuing. For visual effects, you can also distort the appearance of an image by deliberately entering height and width information that differs from the image's actual dimensions.

Spacing Around an Image

If you want a little padding between an image and surrounding text or images, specify blank space for either the left and right sides or the top and bottom sides. You cannot (yet) put different amounts of spacing on, say, the left and right sides.

You can also specify an amount of additional padding surrounding the entire image, called the border. The border puts a uniform amount of spacing on all four sides of the image. Normally invisible, a border will appear with link coloring if the image is used as a hyperlink. You can set the border to zero (0) to hide this hyperlink giveaway (if you believe it mars the appearance of the page, for example) in Browser View. However, in Netscape Composer, the border will appear no matter what, to remind you that the image is a link.

Where to Put the Image?

There are two different methods you can use for storing your images. The easiest method is to keep all images used on a given page within the same folder (or directory) as the page itself. This works best unless you have certain images that are used in many different pages. If you do, then it does not make sense to keep a separate copy of such an image in every folder where it might appear.

By default, an option called "Leave image at the original location" is not checked. This brings any image you insert onto the page into the same folder as the document. Check this option if you have already placed the image in a different folder, as long as it retains the same relative relationship to the current folder that it will have when published on the server.

Actually Inserting the Image

When you're done choosing all the options for your image, click OK. The image will be inserted into your page.

EXPERT ADVICE

If you're comfortable doing it, you can drag and drop images from Windows Explorer straight into a Web page.

Changing How Graphics Are Displayed

Once you've got an image inserted into your document, it can be moved around just as you move text around in the flow of paragraphs. You can select an image, cut it, and paste it elsewhere.

If you want to change any of the display features (such as the alignment of the figure, its size, the spacing around it, alternate text, and so on), just double-click the image—or right-click on the image (click and hold with a Mac) and then select Image Properties from the menu that pops up.

This brings up the Properties dialog box described in the previous few sections.

I Break for Images

To control the way images and text stack up, you might sometimes want to insert a break after an image. To do this, select Insert | Break Below Image(s).

Multimedia: Will Your Audience Care?

We'll cut right to the chase: you can add all kinds of multimedia bells and whistles, but odds are your readers won't turn to your site because of them. You can insert just about any kind of multimedia, from sounds to animations to video, into your Web pages. But they can make pages load too slowly unless your readers are viewing your pages over a fast connection—such as their company's intranet.

Netscape Composer does not help you create any multimedia elements in your Web pages, so we won't cover this in any more depth. For more information about making fancy-pants multimedia elements in Web pages using other tools, you can turn to other books such as (warning: shameless plug for the publisher of this book) *Beyond HTML* by Richard Karpinski (Osborne/McGraw-Hill, 1996), *The World Wide Web Complete Reference* by Rick Stout (Osborne/McGraw-Hill, 1996), and, most of all, *Multimedia: Making It Work, Third Edition* by Tay Vaughan (Osborne/McGraw-Hill, 1996).

Realistically, you're reading this book because you're a busy person (remember the title?). So rather than delve into more advanced topics, let's move on. Remember to see Appendix A for links to Web sites that can tell you more about adding multimedia to your pages.

Creating Forms

If you've ever typed anything on a Web page, you've probably used an HTML-based form. Forms take their name from fill-in-the-blank paper forms—they're ways for Web site designers to get information *back* from Web site visitors.

Regrettably, Netscape Composer doesn't make creating forms as easy as it does other formatting. We'll run you through the HTML coding that makes a form look good on the screen. Although it's a little tedious, it's not too hard, and if you've become accustomed to working with dialog boxes, then you'll recognize the different types of doodads that make up a form. However, it's another thing entirely to get the forms working on the "back end" (the server side) so that when your reader fills out a form, something actually happens. Finally, it's yet another matter to actually entice any of your visitors to fill out the form. The latter is an issue for the marketers and the psychologists.

But what can you do with forms? Any time you want to solicit input from a reader and then respond to what's typed, a form is the way to go. Popular uses include order forms (for online commerce), search forms (for searching the site or the entire Net), and mail forms (for directing feedback into a database or archive).

Constructing a Form in a Document

Forms begin with <FORM> and end with </FORM>. The FORM tag includes two additional attributes: METHOD=, which can be set to GET or POST (it depends on the server—use GET if you're not sure); and ACTION= "*URL of script for form*".

Just before the end of the form, it's customary to include two buttons: one for submitting the filled out form, and one for clearing the form and starting over. The tags for each, respectively, are <INPUT TYPE="submit" VALUE="*optional text, instead of 'Submit'*" *<INPUT TYPE="reset" VALUE="optional text, instead of 'Reset'*".

In the middle go the tags for the form itself—that is, the various blanks and choices that make up the form. The tags for form-entry include the following:

- <TEXTAREA NAME="*name*" ROWS=*x* COLS=*y*>*optional suggested default text* </TEXTAREA> for text boxes
- <INPUT NAME="*name*" TYPE="*text, password, check box, radio, int, hidden, submit, reset*" VALUE="*default value, if any* " SIZE="*optional size*" MAXLENGTH="*optional maximum length*"> for the various input types available (Note: passwords appear as asterisks when entered; multiple check boxes can be checked at once; include the tag CHECKED to have a check box checked by default; only one radio button can be selected at a time; *int* stands for integer, whole number; *hidden* is used for information that gets sent with the form no matter what.)

- <SELECT NAME="*name*" SIZE="*optional number of items to be shown at a time*"> and </SELECT> for drop-down lists (described next)

The SELECT tag produces a drop-down list. To create items for the list, include them between the SELECT tags, starting each one with <OPTION>. To assign a default option to a list, use <OPTION SELECTED> for that one instead. Add MULTIPLE to the SELECT tag to allow more than one item from the list to be selected at once.

Figure 12.5 shows a form utilizing many of the input options.

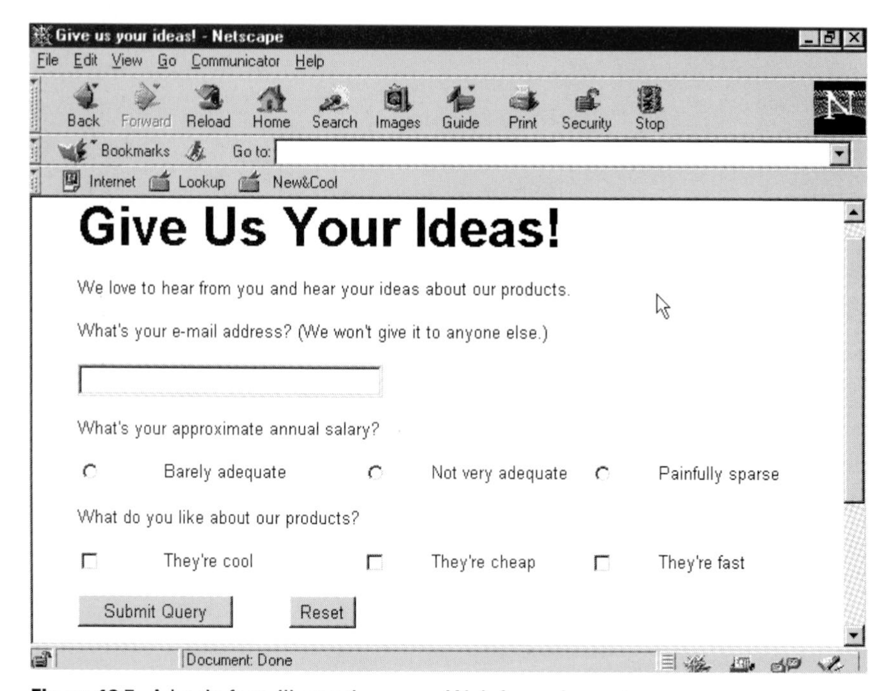

Figure 12.5 A basic form illustrating many Web form elements

To make your form do anything, you'll need to write or borrow a script and put it, or have it put, on your server. Still, even if you put the *functionality* of your form in the hands of a programmer, the appearance, layout, order, and structure of your form is a design issue that you should consider as carefully as you would the rest of your home page.

Writing, Stealing, or Hiring Scripts

Both traditional server-side image maps and forms require the cooperation of a server to function properly. The ingredient that tells the server what to do, how to do it, and when to do it is the CGI script.

The most common type of CGI script is the type that responds to submitted form data. The script file, containing the actual text instructions, is the file referred to in the ACTION="*URL*" part of the <FORM> tag (discussed earlier).

For a query form that's connected to a database, for example, the following things happen when the user clicks the Submit button:

1. The browser sends the request entered into the form to the server, which starts the specified script.
2. The script carries the request to the database manager program (this is why it's called a *gateway* script, because it negotiates between the server and other functions of the host computer or network) and submits a query in a format the database understands.
3. The database returns a result to the script and the script then formats it into an HTML document, which it passes back to the server.
4. The server sends the newly created Web document to the browser, which displays it as the next page the reader sees.

Scripts can actually be written in any language that the server computer understands. They can even be, although they rarely are, full-fledged compiled programs.

Clearly, a thorough discussion of CGI scripting could fill a whole book (and has already filled many). For the most common or the most popular uses of scripts, you can probably find an existing script that works. More than likely, though, you'll have to adapt it slightly to the particular configuration details of your server and the machine it runs on (that's the rub). Probably your best bet is to hire or borrow a CGI guru to get your script to a custom fit.

Sophisticated Navigation Tricks

Besides the basic technique for helping readers navigate—supplying each page with consistently organized hyperlinks—there are a few newer kinds of guides. The first, the image map, can send a reader to different locations based on where they click within a graphic. The second, frames, just divide up Web pages into independent sections (like a TV dinner). They designate some sections for certain purposes (for navigation, banners, headers), while reserving some for other reasons (such as displaying the primary content of the site). Third and fourth are JavaScript and dynamic style sheets—new, complex ways to create dynamic pages.

Image Maps and Three Ways to Make Them

Most image maps are graphical versions of what you might call a key-word index. They often show some form of illustration with formatted text on top. Depending on what word or button you click, you will be sent to a different destination. That's the typical sort of image map, but it's not the only type. Image maps can also be used creatively to supply a graphical depiction of a place with a number of different links. You might choose to use an image map so that you'll have more control over the appearance of your Web page—your organization and links can have a consistent appearance in whatever font you like. See Figure 12.6 for a sample image map.

Most of the time, you'll use maps to lead readers to the major categories of pages at your site. The first step then, in the creation of any image map, is the production of the artwork. If you want something as basic as plain formatted text or a simple graphic, you can assemble the image yourself. If you want something more sophisticated, you may need to hire a graphic artist.

Two Kinds of Image Maps

The oldest type of image maps, called server-side image maps, rely on Web server capabilities to process clicks inside a graphic. Realistically, you'll find these kinds of image maps are rarely used and somewhat complex to set up (on the server side, anyway).

Instead, we'll spend a moment showing you how to set up the second and most common type of image map: client-side image maps.

New Client-Side Maps

It's silly, really, to go to the server to figure out the meaning of an image map click, when the browser could do the job just as well and faster. These maps

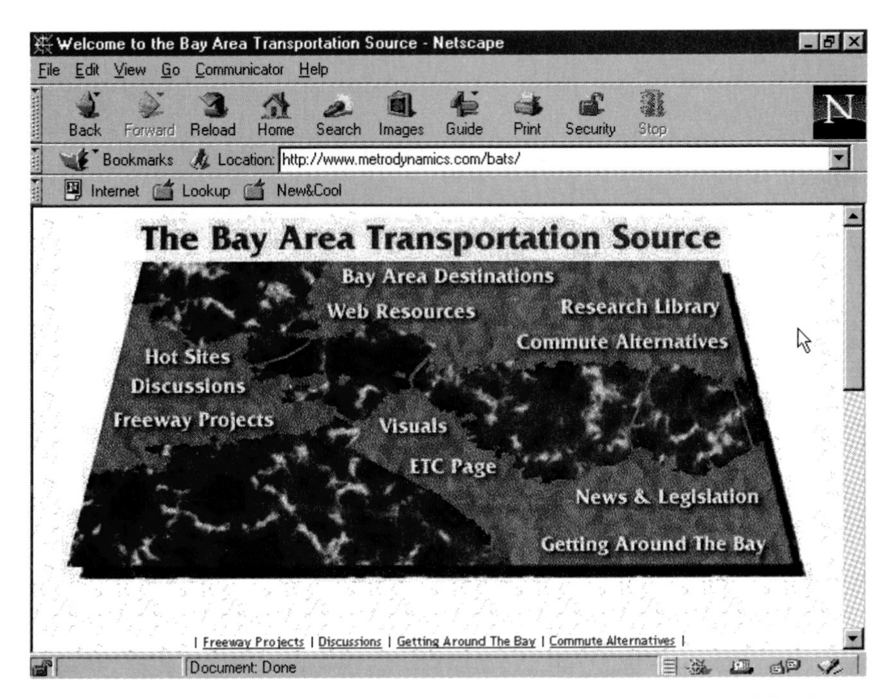

Figure 12.6 This image map creatively superimposes text links on a map of the San Francisco Bay Area

are called client-side image maps because the "thinking" is done on the client, or browser, side of the transaction instead of on the server end. Netscape and Microsoft have incorporated client-side image map recognition into Navigator and Internet Explorer, so this type of map is fairly safe to use. It also provides for alternate text for each region of the map.

Again, you're reading this book because you're busy. We could show you all the individual HTML tags that go into making these kinds of image maps. Or, you could save a lot of time and frustration by using one of the ubiquitous shareware tools available to create image maps. We've used MapTHIS!, among others (see Appendix A for more information and download sites). But don't buy another product if you don't need to: if you own Corel DRAW 7 or Corel WordPerfect Suite 8 (with Corel Presentations 8), for example, these programs already know how to create image maps.

EXPERT ADVICE

You can also make a quick and dirty image map just by slicing up a graphic into pieces and then inserting them side by side into your document. Make sure each image-piece has no border—and no horizontal spacing, for that matter—and make each piece a link to the destination it corresponds to. Add alternate text as you would to any image.

Frames: The High-Tech Look

Frames were introduced by Netscape. Ironically, however, Netscape Composer doesn't help you make them. However, if you're using a more sophisticated Web page design tool (such as Adobe PageMill or Corel WebDesigner), you can make them automatically. We'll tell you about them here, but we won't delve into the arcane HTML code required to create them. Realistically, if you want to make frames, you'll spend less time going to the store and buying a more powerful page creation tool than you will trying to fine-tune the HTML code.

EXPERT ADVICE

If you still feel like you need to learn how to create frames using HTML code, check out the sites listed in Appendix A for more information on HTML coding.

If you've ever worked with a spreadsheet program that allowed the screen to be broken up into panes, or if you've ever split the screen of a word processor, you already have an inkling of what frames look like and do on Web pages. (The Web page shown in Figure 12.4 contains frames.) Frames are subdivided regions of the screen. Each frame can behave more or less as a separate Web document, so it's not just a fancy type of table! Frames can also contain links that point into other frames, such that clicking in one frame (a navigation area across one side of the screen) can result in a change in another frame (perhaps in the large, primary area). Figure 12.4 shows a Web site that uses frames in its high-end, "full service" area.

The problems with frames can range from the minor to the severe. Truth be told, not all browsers recognize them (but frames have a built-in way of offering substitute pages to browsers that don't "do" frames). Compounding this is the nuisance that the standards are still in flux, with, typically, Netscape and Microsoft tossing around changes. And finally, there are the more fundamental issues of whether they're worth the bother or if they're just plain ugly (not to mention wasteful of precious screen real estate!).

Surely, in the long run, some method for keeping fixed navigation tools or other features on the screen while allowing other sections to change will be widely adopted. And frames might just be the way to do it—but that's not yet perfectly clear.

The ultimate problem with frames is that the coding can be quite tricky, depending on what behavior you want, and poorly coded frame pages can cause unreadable pages that may even crash a user's browser!

Dabbling with JavaScript

The final way to jazz up Web documents—one that's within reach of a normal busy person—is to incorporate JavaScript commands into your Web document. JavaScript is not the same thing as Java itself. Nor is it the same thing as the CGI scripts described in the previous section. Java is a much more fully functional programming language, and CGI scripts run on the server side. JavaScript is interpreted by the browser (on the client side).

As with CGI scripting, it's not really possible to learn the ins and outs of JavaScript casually. However, there are a number of useful existing JavaScript routines that you can borrow and adapt for your own pages, the same way you might borrow and adapt someone else's HTML code.

Examples of popular, simple JavaScript routines include a scrolling (possibly annoying to some) "marquee" message in the browser's status bar and a script using a command called "onMouse Over" that puts explanatory text instead of the destination URL in the status bar when the user positions the mouse pointer over a link.

Your next move is to create some more Web pages! As you learn to create more complex Web pages, you'll be ready to take the next step, learning more HTML code or trying an industrial-strength Web page creation tool. In short, what you need to do next is to put down this book and give all your new knowledge a try!

APPENDIX

A 101 Hot Web Sites for Busy People

This list of cool Web sites isn't going to be exactly like every other list you see. What you'll find here is a mix of our favorites and sites that we think you'll find useful. Use our list of favorites to branch out and find your own favorites; follow the "useful" links to learn more about topics discussed in this book. You'll find a wealth of information on the Web—this list will help you find the cream of it.

Sites We Visit Every Day

Here are a few sites we find worth stopping in on—just like reading the morning paper.

1. Good Morning Silicon Valley (**www.gmsv.com**): A Mercury News service that summarizes the day's technology news.
2. New York Times (**www.nytimes.com**): The "paper of record" that includes a newly revitalized CyberTimes section.
3. CNET's News.com (**www.news.com**): A Web-only, technology news summary.
4. USA Today (**www.usatoday.com**): For "McNews" in McNugget-sized bites. Good weather pages.
5. ZDNet's AnchorDesk (**www.anchordesk.com**): A computer industry news site helmed by ex-*Windows Watcher* editor and astute industry observer Jesse Berst.
6. Fortune Business Report (**www.pathfinder.com/fortune/fbr**): A daily report and stock market summary from Jeff's favorite business magazine.

Sites We Visit a Lot

Some sites don't warrant a daily visit, but perhaps a visit every week is still worthwhile.

1. Jazz Central Station (**www.jazzcentralstation.com**): A good starting point for jazz-related clips, news, and music.
2. Business Week Online (**www.businessweek.com**): Part of TimeWarner's monolithic Pathfinder site (which, apparently, is about to become less so).
3. Fast Company (**www.fastcompany.com**): Harbors an active community based on the new business magazine.
4. Fortune Magazine (**www.fortune.com**): Includes material that's not available in the print version.

5. Upside Magazine (**www.upside.com**): Billed as a magazine for the "technology elite," it includes often irreverent commentary on computer industry happenings and trends.
6. Red Herring (**www.herring.com**): In many ways Upside's closest competitor; focuses more on the financial side of the technology business.
7. Cowles Media Central (**www.mediacentral.com**): Brings together Cowles publishing's diverse media insider publications to give insights and reports on media trends.
8. The Drudge Report (**www.drudgereport.com**): Includes controversial Web-based columnist Matt Drudge's writings, but lists links for a quick hub for news and opinion on the Web.
9. Sean's Comics Page (**ape.com/outthere/**): Highly amusing, ballpoint-drawn comics by a multitalented barista at Palo Alto's Café Verona.
10. Martha Stewart (**www.marthastewart.com**) While you can say what you will about her ubiquitousness, the new Martha Stewart Living site has some useful tips for real life and homeownership.

Stuff for Real Life and Home

But not everything we visit has to do with news or computers—we have real lives, too.

1. This Old House (**www.thisoldhouse.com**): Great resources from the magazine spinoff of the television series to help you remodel or maintain your house.
2. Readers Digest Association (**www.readersdigest.com**): Includes material from *American Handyman* magazine, with information on home projects, gardening, recipes, and more.
3. Better Homes and Gardens (**www.bhglive.com**): Includes useful tips on decorating and just around-the-house projects and tips.

Kids and Parents

What's keeping Jeff from getting those home projects done are his kids—so he turns to the Web every now and again for a little parenting advice.

1. Discovery Channel Online (**www.discovery.com**): A Web pioneer, it continues to offer family-friendly, educational information that makes you glad you don't live in the tundra (and if you do, no offense meant).
2. Family.com (**www.family.com**): Part of the monstrous Disney.com site, offering fun and information for parents and kids alike.
3. Parent Soup (**www.parentsoup.com**): Offers an online community to help parents realize they're not alone in the challenges they face.
4. ParentTime (**www.parenttime.com**): Sports in-depth articles on successfully living with children—and surviving.

Reference Works

Why have a shelf full of reference books when you've got the essentials on the Web?

1. The Elements of Style (**www.cs.williams.edu/~bailey/style/**): An instant-access, electronic copy of that dog-eared guide to clear communication.
2. Writers Write (**www.writerswrite.com/journal/**): An online journal for those who write.
3. Jack Lynch's Grammar and Style Notes (**www.english.upenn.edu/~jlynch/grammar.html**): A nice summary of good English.
4. Style FAQ (**www.rt66.com/~telp/sfindex.htm**): A style guide created by the members of the Copyediting mailing list.

Job Hunting

Most newspapers put their classifieds online—and many nationwide resources are available for the job hunter or career switcher, too.

1. Monster Board (**www.monster.com**): A nationwide resource for job listings and job information.
2. CareerPath (**www.careerpath.com**): Great job listings and content.
3. CareerBuilder (**www.careerbuilder.com**): Heavy on advice and resources for people looking to climb the career ladder.
4. CareerMosaic (**www.careermosaic.com**): Another good job hunter's site.

House Hunting

Trust us on these: Jeff just moved to a new state!

1. Realtor.com (**www.realtor.com**): A nationwide gathering of real estate listings and advice (by professional realtors).
2. Homes.com (**www.homes.com**): A home listing site that includes good information on loans, mortgages, and other financial issues.
3. Coldwell Banker online (**www.coldwellbanker.com**): Listings maintained by this nationwide real estate firm.

Just Plain Fun

You just don't want to be productive all the time. The Web's a great time-waster (especially on a slow modem connection).

1. Disney.com (**www.disney.com**): A mega-site encompassing Disney's many properties (movies, theme parks, etc.) is great for any Disney fan, young or old. (Disney's pay-per-month site Daily Blast (**www.dailyblast.com**) has great kids' games.)
2. Addicted to Noise (**www.addict.com**): An online music magazine covering rock in all its forms.
3. David Bowie (**www.davidbowie.com**): A site that's nicely designed and includes nifty stuff from Bowie's latest, "Earthling."
4. BeZerk (**www.bezerk.com**): Berkeley Systems' site dedicated to online games, including the Web version of the smartass trivia game *You Don't Know Jack.*

5. ESPN SportsZone (**espn.sportszone.com**): The premier sports site on the Web, offers stats, news, instant score updates, and a wealth of sports info.
6. Riddler.com (**www.riddler.com**): Offers a wide variety of games and online diversions.
7. Music Newswire (**www.musicnewswire.com**): A compilation of rock-related headlines linked to major music-related sites around the world.

Health and Fitness

Get up from the computer now and then.

1. Stanford HealthLINK (**healthlink.stanford.edu**): A health news site designed for Stanford's community, but it includes information suitable for anyone in the world.
2. Thrive (**www.thriveonline.com**): A wide-ranging health-related site that draws on the expertise of major magazines.

E-mail and Online Etiquette

There's nothing more embarrassing than the social stigma of an e-mail faux pas—short of, perhaps, a dryer sheet stuck to your back or something. Here's help.

1. The Smiley Dictionary (**www.netsurf.org/~violet/Smileys/**): A unique list of smileys—ASCII ways to express emotion.
2. Claris guide to email etiquette (**www.claris.com/products/claris/emailer/eguide/index.html**): A good guide to fitting in and behaving yourself online.
3. E-mail Etiquette (**www.iwillfollow.com/email.htm**): A surprisingly brief guide to online behavior.

Downloading Files and Shareware

We mentioned earlier in the book how you can download files—here's a list of resources and sites that'll help you find and use freeware, shareware, and more.

1. WinZip (**www.winzip.com**): The long-time favorite Windows-based file compression utility from Nico Mak Computing.
2. PKWare (**www.pkware.com**): The original authors of PKZip, the most widely used PC file compression utility—with a new Windows version.
3. Aladdin Systems, Inc. (**www.aladdinsys.com**): Makers of StuffIt, the most widely used Macintosh file compression utility.
4. download.com (**www.download.com**): CNET's collection of hot, downloadable files and programs.
5. ZDNet Downloads (**www.hotfiles.com**): Ziff-Davis' huge selection of downloads and reviews.

Online HTML References

For more information about HTML, check out these online references:

1. The HTML Language Specification (**developer.netscape.com/platform/html_compilation/index.html**): A guide to the always-changing specification for HTML.
2. Learning HTML (**www.bev.net/computer/htmlhelp/**): A volunteer-built site designed to help you learn fast.
3. The Beginner's Guide to HTML (**www.ncsa.uiuc.edu/demoweb/html-primer.html**): Just like it sounds!
4. HTML Quick Reference (**kuhttp.cc.ukans.edu/lynx_help/HTML_quick.html**): Need a fast refresher on tag syntax? Turn here.
5. Bare Bones HTML (**www.access.digex.net/~werbach/barebone.html**): A fast way to learn only the basics.

6. Composing good HTML (**www.cs.cmu.edu/~tilt/cgh/**): The elements of HTML style.
7. The WEB DESIGNER (**limestone.kosone.com/people/nelson/nl.htm**): A compendium of HTML and Web design tips.
8. A Suite of HTML validation tools (**www.ccs.org/validate/**): To help check your HTML code and pages.

Learning Good Web Design

The best place to learn about Web design is from good designers.

1. Roger Black (**www.rogerblack.com**): Controversial designer Roger Black offers his style rules for Web design.
2. David Siegel's Killer Sites (**www.killersites.com/**): Lists well-designed Web pages.
3. Will-Harris House (**www.will-harris.com**): DTP pioneer and pundit Daniel Will-Harris' compendium of salient design advice.

Some Favorite Home Pages

Some miscellaneous sites that we particularly enjoy:

1. Levi Asher's Literary Kicks Web site (**www.charm.net/~brooklyn**): One of our favorite home-pages-gone-wild. Levi Asher publishes his thoughts and insights about the Beat Generation writers, as well as other literary luminaries he admires. Like Tom Sawyer painting the fence, he just started pushing his writing and art out there and, before long, started getting a lot of e-mail from other people with similar interests. Soon afterward he had all kinds of volunteers contributing articles, checking information, doing interviews, and so on.
2. Internet Scout home page (**scout.cs.wisc.edu**): An ongoing project since the early days of the Web, lists useful and interesting new Web sites, discussion groups, and other Net resources.

Fun, Artsy, and Experimental Pages

Weird and unusual sites aren't uncommon on the Web, but here are some of the best:

1. Lard (**www.lard.com**): For a complete waste of time.
2. Pythonline (**www.pythonline.com**): For the original complete waste of time.

And a few from Antiweb's Arthur (you'll have to visit these yourself to see why he recommends them):

3. Residence (**www.xs4all.nl/~arthur**)
4. Info (**www.xs4all.nl/~bajazzo**)
5. Studio (**www.art.net/~arthur/welcom.html**)
6. Toys (**www.bajazzo.xs4all.nl/posi-web**)
7. Design (**www.bajazzo.com**)
8. Design (**www.kessels.com**)

Shopping

For those who like to shop in their pajamas in the middle of the night, the Web is a 24-hour mall:

1. CDNow (**www.cdnow.com**): Offers a wide selection and competitive prices on CDs, videos, and more—all delivered promptly.
2. QVC (**www.qvc.com**): Offers the same selection found on the cable television shopping network, but without the annoying hosts.
3. Amazon.com (**www.amazon.com**): The premier online bookseller offers literally every book in print at a discount.

Magazines and Newspapers Worth Reading on the Web

Jeff's in publishing, so here are a few he thinks have the right idea as they move to electronic publishing.

1. San Jose Mercury News (**www.mercurycenter.com**): Offers Mercury Center, a great newspaper that places its content online faster than it publishes it. Exclusive, excellent coverage of computing and Silicon Valley issues a specialty.
2. The Gate (**www.sfgate.com**): A blend of information from the *San Francisco Chronicle, San Francisco Examiner,* and SF TV station KRON. Includes video clips and stills.
3. Wired (**www.wired.com**): A family of sites, including **www.hotwired.com** and **www.webmonkey.com**. The magazine for "digerati" continues to reinvent itself. In the meantime, they're providing some great info (while they look for a way to make themselves more profitable).

Travel

Travel planning is a lot easier with these sites at your disposal:

1. WebFlyer (**www.webflyer.com**): The offshoot of longtime frequent traveler expert Randy Petersen's *Frequent Flyer Magazine* and related efforts.
2. Delta Airlines (**www.delta-air.com**) and American Airlines (**www.americanairlines.com**): Offer online frequent flyer account info as well as online ticket reservation. (Note: We keep trying online travel reservation services such as Microsoft's Expedia, but find them inadequate compared to our reliable, knowledgeable travel agent.)
3. City.Net (**www.city.net**): Features travel advice and nifty, pinpoint accuracy maps.

Stocks and Investing

Wall Street is just a click away with these investment sites:

1. E-Trade (**www.etrade.com**): Offers significant discounts on stock trading, free of commissions and broker fees.
2. Data Broadcasting Corp.'s Newsroom (**www.dbc.com**): Click DBC news from the home page for a frequently updated summary of investing news and leading market indicators.

Computer Jargon Translated

Are you jargon-impaired? Check out these sites to keep up with the vocabulary of the nerds:

1. The Jargon File (**www.ccil.org/jargon/jargon_toc.html**): A listing of both hacker slang and computer jargon (plus some hacker lifestyle insights).
2. Cool Jargon of the Day (**www.bitech.com/jargon/cool**): Helps you build your nerd vocabulary with a new, arcane listing daily.
3. Jargon Watch (**www.wired.com/hardwired/jargonwatch**): The companion Web site to the Jargon Watch book, published by *Wired Magazine's* now defunct book division.
4. Chatter's Jargon Dictionary (**www.tiac.net/users/scg/ jargpge.htm#Jargon**): A comprehensive listing of online chatter's shorthand, smileys, and more.

HTML Converters

Look to these sites to make your HTML experience a little easier:

1. The W3 Organization's page (**www.w3.org/hypertext/WWW/Tools/Filters.html**): Lists scads of useful file converters.
2. The Web Designer's page of HTML Converters and Templates (**limestone.kosone.com/people/nelsonl/convert.htm**): Also contains excellent listings of file format conversion programs.

Resources for Multimedia Pages

HTML's not the only format on the Web. Here are some other common file formats to present Web information.

1. Adobe Acrobat (**www.adobe.com/acrobat**): A powerful tool to publish documents online with formatting, graphics, fonts, etc. intact.
2. RealAudio (**www.real.com**): A way to both get and publish audio and video from/to Web sites in "real time."
3. Macromedia Shockwave (**www.macromedia.com**): A popular tool for adding Web-based animation and multimedia.

Resources for Image Map Creation Tools

You can find shareware and other tools for image map-making at these sites:

1. Mediatec (**www.mediatec.com**): A company who made the formerly ubiquitous MapTHIS! and now promotes LiveImage, a Win 95/Win NT 4.0, client-side image mapping shareware package.
2. Boutell.com (**www.boutell.com/mapedit**): Offers the shareware MapEdit to help you create image maps quickly.
3. Corel (**www.corel.com**): Offers a mess of products that can create image maps in a snap. Our favorite: Corel Xara 1.5, an all-purpose, fast image editing program with image map creation abilities. However, look at CorelDraw and Corel Presentations, part of Corel WordPerfect Suite, too.

Resources for Creating Frames in HTML

Confused about frames? Check these out for some expert guidance:

1. Sharky's Netscape Frames Tutorial (**www.newbie.net/frames**): Offers an easy-to-understand frames tutorial.
2. Frames: An Introduction (**www.netscape.com/assist/net_sites/frames.html**): Netscape's stilted but useful guide to creating frames.

Shameless Self-publicity

And last but certainly not least, some of our other favorite sites—for obvious reasons:

1. The Development Exchange (**www.windx.com** or **www.devx.com**): Fawcette Technical Publications' one-stop shop for software developers.
2. EnterZone (**ezone.org/ez**): Christian's groundbreaking hypertext Web zine.
3. Coffeehousebook.com (**www.coffeehousebook.com**): Christian's other site, dedicated to finding literary talent on the Web (and kind of to promote the book *Coffeehouse: Writings from the Web*, but it has its own content as well).
4. WordPerfect Universe (**www.wpwin.com**): An unofficial site compiling all kinds of in-depth WordPerfect info by WP expert Gordon McComb.

CAUTION

Index

NOTE: Page numbers in *italics* refer to illustrations or charts.

J

K

L

M

N

O

P

Q

R